Tucholsky Wagner Zola Scott Sydow Freud Schlegel
Turgenev Fonatne
Wallace Walther von der Vogelweide Fouqué Friedrich II. von Preußen
Twain Weber Freiligrath Frey
Fechner Fichte Weiße Rose von Fallersleben Kant Ernst Richthofen Frommel
Engels Fielding Hölderlin Tacitus Dumas
Fehrs Faber Flaubert Eichendorff Ebner Eschenbach
Feuerbach Maximilian I. von Habsburg Fock Eliasberg Zweig Vergil
Goethe Ewald Eliot London
Mendelssohn Balzac Shakespeare Elisabeth von Österreich Dostojewski Ganghofer
Trackl Lichtenberg Rathenau Doyle Gjellerup
Mommsen Stevenson Tolstoi Hambruch
Thoma Lenz Hanrieder Droste-Hülshoff
Dach Verne von Arnim Hägele Hauff Humboldt
Karrillon Reuter Rousseau Hagen Hauptmann Gautier
Garschin Defoe Hebbel Baudelaire
Damaschke Descartes Hegel Kussmaul Herder
Wolfram von Eschenbach Dickens Schopenhauer Rilke George
Bronner Darwin Melville Grimm Jerome Bebel Proust
Campe Horváth Aristoteles Voltaire Federer Herodot
Bismarck Vigny Barlach Heine
Gengenbach
Storm Casanova Lessing Tersteegen Gilm Grillparzer Georgy
Chamberlain Langbein Gryphius
Brentano Lafontaine
Strachwitz Claudius Schiller Kralik Iffland Sokrates
Bellamy Schilling
Katharina II. von Rußland Gerstäcker Raabe Gibbon Tschechow
Löns Hesse Hoffmann Gogol Wilde Gleim Vulpius
Luther Heym Hofmannsthal Klee Hölty Morgenstern Goedicke
Roth Heyse Klopstock Kleist
Luxemburg Puschkin Homer Mörike Musil
Machiavelli La Roche Horaz
Navarra Aurel Musset Kierkegaard Kraft Kraus
Nestroy Marie de France Lamprecht Kind Kirchhoff Hugo Moltke
Laotse Ipsen Liebknecht
Nietzsche Nansen Ringelnatz
Marx Lassalle Gorki Klett
von Ossietzky May vom Stein Lawrence Leibniz
Petalozzi Knigge Irving
Platon Pückler Michelangelo Kafka
Sachs Poe Liebermann Kock
de Sade Praetorius Mistral Zetkin Korolenko

The publishing house tredition has created the series **TREDITION CLASSICS**. It contains classical literature works from over two thousand years. Most of these titles have been out of print and off the bookstore shelves for decades.

The book series is intended to preserve the cultural legacy and to promote the timeless works of classical literature. As a reader of a **TREDITION CLASSICS** book, the reader supports the mission to save many of the amazing works of world literature from oblivion.

The symbol of **TREDITION CLASSICS** is Johannes Gutenberg (1400 – 1468), the inventor of movable type printing.

With the series, tredition intends to make thousands of international literature classics available in printed format again – worldwide.

All books are available at book retailers worldwide in paperback and in hardcover. For more information please visit: www.tredition.com

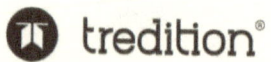

tredition was established in 2006 by Sandra Latusseck and Soenke Schulz. Based in Hamburg, Germany, tredition offers publishing solutions to authors and publishing houses, combined with worldwide distribution of printed and digital book content. tredition is uniquely positioned to enable authors and publishing houses to create books on their own terms and without conventional manufacturing risks.

For more information please visit: www.tredition.com

About Orchids A Chat

Frederick Boyle

Imprint

This book is part of the TREDITION CLASSICS series.

Author: Frederick Boyle
Cover design: toepferschumann, Berlin (Germany)

Publisher: tradition GmbH, Hamburg (Germany)
ISBN: 978-3-8491-8833-7

www.tredition.com
www.tredition.de

Copyright:
The content of this book is sourced from the public domain.

The intention of the TREDITION CLASSICS series is to make world literature in the public domain available in printed format. Literary enthusiasts and organizations worldwide have scanned and digitally edited the original texts. tredition has subsequently formatted and redesigned the content into a modern reading layout. Therefore, we cannot guarantee the exact reproduction of the original format of a particular historic edition. Please also note that no modifications have been made to the spelling, therefore it may differ from the orthography used today.

I INSCRIBE
THIS BOOK TO MY GUIDE, COMFORTER
AND FRIEND,

JOSEPH GODSEFF.

CONTENTS.

My Gardening

An Orchid Sale

Orchids

Cool Orchids

Warm Orchids

Hot Orchids

The Lost Orchid

An Orchid Farm

Orchids and Hybridizing

LIST OF ILLUSTRATIONS.

Vanda Sanderiana

Odontoglossum crispum Alexandræ

Oncidium macranthum

Dendrobium Brymerianum

Cœlogene pandurata

Cattleya labiata

Lœlia anceps Schroederiana

Cypripedium (hybridum) Pollettianum

PREFACE.

The purport of this book is shown in the letter following which I addressed to the editor of the *Daily News* some months ago:—

"I thank you for reminding your readers, by reference to my humble work, that the delight of growing orchids can be enjoyed by persons of very modest fortune. To spread that knowledge is my contribution to philanthropy, and I make bold to say that it ranks as high as some which are commended from pulpits and platforms. For your leader-writer is inexact, though complimentary, in assuming that any 'special genius' enables me to cultivate orchids without more expense than other greenhouse plants entail, or even without a gardener. I am happy to know that scores of worthy gentlemen—ladies too—not more gifted than their neighbours in any sense, find no greater difficulty. If the pleasure of one of these be due to any writings of mine, I have wrought some good in my generation."

With the same hope I have collected those writings, dispersed and buried more or less in periodicals. The articles in this volume are collected—with permission which I gratefully acknowledge—from *The Standard, Saturday Review, St. James's Gazette, National Review*, and *Longman's Magazine*. With some pride I discover, on reading them again, that hardly a statement needs correction, for they contain many statements, and some were published years ago. But in this, as in other lore, a student still gathers facts. The essays have been brought up to date by additions—in especial that upon "Hybridizing," a theme which has not interested the great public hitherto, simply because the great public knows nothing about it. There is not, in fact, so far as I am aware, any general record of the amazing and delightful achievements which have been made therein of late years. It does not fall within my province to frame such a record. But at least any person who reads this unscientific account, not daunted by the title, will understand the fascination of the study.

These essays profess to be no more than chat of a literary man about orchids. They contain a multitude of facts, told in some detail where such attention seems necessary, which can only be found elsewhere in baldest outline if found at all. Everything that relates to orchids has a charm for me, and I have learned to hold it as an article of faith that pursuits which interest one member of the cultured public will interest all, if displayed clearly and pleasantly, in a form to catch attention at the outset. Savants and professionals have kept the delights of orchidology to themselves as yet. They smother them in scientific treatises, or commit them to dry earth burial in gardening books. Very few outsiders suspect that any amusement could be found therein. Orchids are environed by mystery, pierced now and again by a brief announcement that something with an incredible name has been sold for a fabulous number of guineas; which passing glimpse into an unknown world makes it more legendary than before. It is high time such noxious superstitions were dispersed. Surely, I think, this volume will do the good work—if the public will read it.

The illustrations are reduced from those delightful drawings by Mr. Moon admired throughout the world in the pages of "Reichenbachia." The licence to use them is one of many favours for which I am indebted to the proprietors of that stately work.

I do not give detailed instructions for culture. No one could be more firmly convinced that a treatise on that subject is needed, for no one assuredly has learned, by more varied and disastrous experience, to see the omissions of the text-books. They are written for the initiated, though designed for the amateur. Naturally it is so. A man who has been brought up to business can hardly resume the utter ignorance of the neophyte. Unconsciously he will take a certain degree of knowledge for granted, and he will neglect to enforce those elementary principles which are most important of all. Nor is the writer of a gardening book accustomed, as a rule, to marshal his facts in due order, to keep proportion, to assure himself that his directions will be exactly understood by those who know nothing.

The brief hints in "Reichenbachia" are admirable, but one does not cheerfully refer to an authority in folio. Messrs. Veitch's "Manual of Orchidaceous Plants" is a model of lucidity and a mine of information. Repeated editions of Messrs. B.S. Williams' "Orchid Growers' Manual" have proved its merit, and, upon the whole, I have no hesitation in declaring that this is the most useful work which has come under my notice. But they are all adapted for those who have passed the elementary stage.

Thus, if I have introduced few remarks on culture, it is not because I think them needless. The reason may be frankly confessed. I am not sure that my time would be duly paid. If this little book should reach a second edition, I will resume once more the ignorance that was mine eight years ago, and as a fellow-novice tell the unskilled amateur how to grow orchids.

Frederick Boyle.

North Lodge, Addiscombe, 1893.

ABOUT ORCHIDS

MY GARDENING.

I.

The contents of my Bungalow gave material for some "Legends" which perhaps are not yet universally forgotten. I have added few curiosities to the list since that work was published. My days of travel seem to be over; but in quitting that happiest way of life—not willingly—I have had the luck to find another occupation not less interesting, and better suited to grey hairs and stiffened limbs. This volume deals with the appurtenances of my Bungalow, as one may say—the orchid-houses. But a man who has almost forgotten what little knowledge he gathered in youth about English plants does not readily turn to that higher branch of horticulture. More ignorant even than others, he will cherish all the superstitions and illusions which environ the orchid family. En [Pg 2]lightenment is a slow process, and he will make many experiences before perceiving his true bent. How I came to grow orchids will be told in this first article.

The ground at my disposal is a quarter of an acre. From that tiny area deduct the space occupied by my house, and it will be seen that myriads of good people dwelling in the suburbs, whose garden, to put it courteously, is not sung by poets, have as much land as I. The aspect is due north—a grave disadvantage. Upon that side, from the house-wall to the fence, I have forty-five feet, on the east fifty feet, on the south sixty feet, on the west a mere *ruelle*. Almost every one who works out these figures will laugh, and the remainder sneer. Here's a garden to write about! That area might do for a tennis-court or for a general meeting of Mr. Frederic Harrison's persuasion. You might kennel a pack of hounds there, or beat a carpet, or assemble those members of the cultured class who admire Mr. Gladstone. But grow flowers—roses—to cut by the basketful, fruit to make jam for a jam-eating household the year round, mushrooms, tomatoes, water-lilies, orchids; those Indian jugglers who bring a mango-tree to perfection on your verandah in twenty minutes might be able to do it, but not a consistent [Pg 3] Christian. Nevertheless I affirm that I have done all these things, and I shall even venture to make other demands upon the public credulity.

When I first surveyed my garden sixteen years ago, a big Cupressus stood before the front door, in a vast round bed one half of which would yield no flowers at all, and the other half only spindlings. This was encircled by a carriage-drive! A close row of limes, supported by more Cupressus, overhung the palings all round; a dense little shrubbery hid the back door; a weeping-ash, already tall and handsome, stood to eastward. Curiously green and snug was the scene under these conditions, rather like a forest glade; but if the space available be considered and allowance be made for the shadow of all those trees, any tiro can calculate the room left for grass and flowers—and the miserable appearance of both. Beyond that dense little shrubbery the soil was occupied with potatoes mostly, and a big enclosure for hens.

First I dug up the fine Cupressus. They told me such a big tree could not possibly "move;" but it did, and it now fills an out-of-the-way place as usefully as ornamentally. I suppressed the carriage-drive, making a straight path broad enough for pedestrians only, and cut down a number of the trees. The blessed [Pg 4]sunlight recognized my garden once more. Then I rooted out the shrubbery; did away with the fowl-house, using its materials to build two little sheds against the back fence; dug up the potato-garden—made *tabula rasa*, in fact; dismissed my labourers, and considered. I meant to be my own gardener. But already, sixteen years ago, I had a dislike of stooping. To kneel was almost as wearisome. Therefore I adopted the system of raised beds—common enough. Returning home, however, after a year's absence, I found my oak posts decaying—unseasoned, doubtless, when put in. To prevent trouble of this sort in future, I substituted drain-pipes set on end; the first of those ideas which have won commendation from great authorities. Drainpipes do not encourage insects. Filled with earth, each bears a showy plant—lobelia, pyrethrum, saxifrage, or what not, with the utmost neatness, making a border; and they last eternally. But there was still much stooping, of course, whilst I became more impatient of it. One day a remedy flashed through my mind: that happy thought which became the essence or principle of my gardening, and makes this account thereof worth attention perhaps. Why not raise to a comfortable level all parts of the area over [Pg 5]which I had need to bend? Though no horticulturist, perhaps, ever had such a thought before, expense was the sole objection visible. Called away just then for another long absence, I gave orders that no "dust" should leave the house; and found a monstrous heap on my return. The road-contractors supplied "sweepings" at a shilling a load. Beginning at the outskirts of my property, I raised a mound three feet high and three feet broad, replanted the shrubs on the back edge, and left a handsome border for flowers. So well this succeeded, so admirably every plant throve in that compost, naturally drained and lifted to the sunlight, that I enlarged my views.

The soil is gravel, peculiarly bad for roses; and at no distant day my garden was a swamp, not unchronicled had we room to dwell on such matters. The bit of lawn looked decent only at midsummer. I first tackled the rose question. The bushes and standards, such as they were, faced south, of course—that is, behind the house. A line of fruit-trees there began to shade them grievously. Experts assured me that if I raised a bank against these, of such a height as I proposed, they would surely die; I paid no attention to the experts, nor did my fruit-trees. The mound raised is, in fact, a crescent on the inner edge, [Pg 6]thirty feet broad, seventy feet between the horns, square at the back behind the fruit-trees; a walk runs there, between it and the fence, and in the narrow space on either hand I grow such herbs as one cannot easily buy—chervil, chives, tarragon. Also I have beds of celeriac, and cold frames which yield a few cucumbers in the summer when emptied of plants. Not one inch of ground is lost in my garden.

The roses occupy this crescent. After sinking to its utmost now, the bank stands two feet six inches above the gravel path. At that elevation they defied the shadow for years, and for the most part they will continue to do so as long as I feel any interest in their well-being. But there is a space, the least important fortunately, where the shade, growing year by year, has got the mastery. That space I have surrendered frankly, covering it over with the charming saxifrage, *S. hypnoides*, through which in spring push bluebells, primroses, and miscellaneous bulbs, while the exquisite green carpet frames pots of scarlet geranium and such bright flowers, movable at will. That saxifrage, indeed, is one of my happiest devices. Finding that grass would not thrive upon the steep bank of my mounds, I dotted them over with tufts of it, which have spread, until at this [Pg 7]time they are clothed in vivid green the year round, and white as an untouched snowdrift in spring. Thus also the foot-wide paths of my rose-beds are edged; and a neater or a lovelier border could not be imagined.

With such a tiny space of ground the choice of roses is very important. Hybrids take up too much room for general service. One must have a few for colour; but the mass should be Teas, Noisettes, and, above all, Bengals. This day, the second week in October, I can pick fifty roses; and I expect to do so every morning till the end of the month in a sunny autumn. They will be mostly Bengals; but there are two exquisite varieties sold by Messrs. Paul—I forget which of them—nearly as free flowering. These are Camoens and Mad. J. Messimy. They have a tint unlike any other rose; they grow strongly for their class, and the bloom is singularly graceful.

The tiny but vexatious lawn was next attacked. I stripped off the turf, planted drain-pipes along the gravel walk, filled in with road-sweepings to the level of their tops, and relaid the turf. It is now a little picture of a lawn. Each drain-pipe was planted with a cutting of ivy, which now form a beautiful evergreen roll beside the path. Thus [Pg 8]as you walk in my garden, everywhere the ground is more or less above its natural level; raised so high here and there that you cannot look over the plants which crown the summit. Any gardener at least will understand how luxuriantly everything grows and flowers under such conditions. Enthusiastic visitors declare that I have "scenery," and picturesque effects, and delightful surprises, in my quarter-acre of ground! Certainly I have flowers almost enough, and fruit, and perfect seclusion also. Though there are houses all round within a few yards, you catch but a glimpse of them at certain points while the trees are still clothed. Those mounds are all the secret.

II.

I was my own gardener, and sixteen years ago I knew nothing whatever of the business. The process of education was almost as amusing as expensive; but that fashion of humour is threadbare. In those early days I would have none of your geraniums, hardy perennials, and such common things. Diligently studying the "growers'" [Pg 9] catalogues, I looked out, not novelties alone, but curious novelties. Not one of them "did any good" to the best of my recollection. Impatient and disgusted, I formed several extraordinary projects to evade my ignorance of horticulture. Among others which I recollect was an idea of growing bulbs the year round! No trouble with bulbs! you just plant them and they do their duty. A patient friend at Kew made me a list of genera and species which, if all went well, should flower in succession. But there was a woeful gap about midsummer—just the time when gardens ought to be brightest. Still, I resolved to carry out the scheme, so far as it went, and forwarded my list to Covent Garden for an estimate of the expense. It amounted to some hundreds of pounds. So that notion fell through.

But the patient friend suggested something for which I still cherish his memory. He pointed out that bulbs look very formal mostly, unless planted in great quantities, as may be done with the cheap sorts—tulips and such. An undergrowth of low brightly-coloured annuals would correct this disadvantage. I caught the hint, and I profit by it to this more enlightened day. Spring bulbs are still a *spécialité* of my gardening. I buy them fresh every autumn—but of Messrs. [Pg 10] Protheroe and Morris, in Cheapside; not at the dealers'. Thus they are comparatively inexpensive. After planting my tulips, narcissus, and such tall things, however, I clothe the beds with forget-me-not or *Silene pendula*, or both, which keep them green through the winter and form a dense carpet in spring. Through it the bulbs push, and both flower at the same time. Thus my brilliant tulips, snowy narcissus poeticus, golden daffodils, rise above and among a sheet of blue or pink—one or the other to match their hue—and look infinitely more beautiful on that ground colour. I venture to say, indeed, that no garden on earth can be more lovely than mine while the forget-me-not and the bulbs are flowering together. This may be a familiar practice, but I never met with it elsewhere.

Another wild scheme I recollect. Water-plants need no attention. The most skilful horticulturist cannot improve, the most ignorant cannot harm them. I seriously proposed to convert my lawn into a tank two feet deep lined with Roman cement and warmed by a furnace, there to grow tropical nymphæa, with a vague "et cetera." The idea was not so absolutely mad as the unlearned may think, for two of my relatives were first and second to flower *Victoria Regia* in the open-air—but they [Pg 11]had more than a few feet of garden. The chances go, in fact, that it would have been carried through had I been certain of remaining in England for the time necessary. Meanwhile I constructed two big tanks of wood lined with sheet-zinc, and a small one to stand on legs. The experts were much amused. Neither fish nor plant, they said, could live in a zinc vessel. They proved to be right in the former case, but utterly wrong in the latter—which, you will observe, is their special domain. I grew all manner of hardy nymphæa and aquatics for years, until my big tanks sprung a leak. Having learned by that time the ABC, at least, of *terra-firma* gardening, I did not trouble to have them mended. On the contrary, making more holes, I filled the centre with Pampas grass and variegated Eulalias, set lady-grass and others round, and bordered the whole with lobelia—renewing, in fact, somewhat of the spring effect. Next year, however, I shall plant them with *Anomatheca cruenta*—quaintest of flowering grasses, if a grass it must be called. This charming species from South Africa is very little known; readers who take the hint will be grateful to me. They will find it decidedly expensive bought by the plant, as growers prefer to sell. But, with a little pressing seed may be obtained, and it [Pg 12]multiplies fast. I find *Anomatheca cruenta* hardy in my sheltered garden.

The small tank on legs still remains, and I cut a few *Nymphæa odorata* every year. But it is mostly given up to *Aponogeton distachyon*—the "Cape lily." They seed very freely in the open; and if this tank lay in the ground, long since their exquisite white flowers, so strange in shape and so powerful of scent, would have stood as thick as blades of grass upon it—such a lovely sight as was beheld in the garden of the late Mr. Harrison, at Shortlands. But being raised two feet or so, with a current of air beneath, its contents are frozen to a solid block, soil and all, again and again, each winter. That a Cape plant should survive such treatment seems incredible—contrary to all the books. But my established Aponogeton do somehow; only the seedlings perish. Here again is a useful hint, I trust. But evidently it would be better, if convenient, to take the bulbs indoors before frost sets in.

Having water thus at hand, it very soon occurred to me to make war upon the slugs by propagating their natural enemies. Those banks and borders of *Saxifraga hypnoides*, to which I referred formerly, exact some precaution of the kind. Much as every one who sees admires them, the slugs, no [Pg 13]doubt, are more enthusiastic still. Therefore I do not recommend that idea, unless it be supplemented by some effective method of combating a grave disadvantage. My own may not commend itself to every one. Each spring I entrust some casual little boy with a pail; he brings it back full of frog-spawn and receives sixpence. I speculate sometimes with complacency how many thousand of healthy and industrious batrachians I have reared and turned out for the benefit of my neighbours. Enough perhaps, but certainly no more, remain to serve me—that I know because the slugs give very little trouble in spite of the most favourable circumstances. You can always find frogs in my garden by looking for them, but of the thousands hatched every year, ninety-nine per cent. must vanish. Do blackbirds and thrushes eat young frogs? They are strangely abundant with me. But those who cultivate tadpoles must look over the breeding-pond from time to time. My whole batch was devoured one year by "devils"—the larvæ of *Dytiscus marginalis*, the Plunger beetle. I have benefited, or at least have puzzled my neighbours also by introducing to them another sort of frog. Three years ago I bought twenty-five Hylœ, the pretty green tree species, to dwell in my Odontoglossum house and [Pg 14]exterminate the insects. Every ventilator there is covered with perforated zinc—to prevent insects getting in; but, by some means approaching the miraculous, all my Hylœ contrived to escape. Several were caught in the garden and put back, but again they found their way to the open-air; and presently my fruit-trees became vocal. So far, this is the experience of every one, probably, who has tried to keep green frogs. But in my case they survived two winters—one which everybody recollects, the most severe of this generation. My frogs sang merrily through the summer; but all in a neighbour's garden. I am not acquainted with that family; but it is cheering to think how much innocent diversion I have provided for its members.

Pleasant also it is, by the way, to vindicate the character of green frogs. I never heard them spoken of by gardeners but with contempt. Not only do they persist in escaping; more than that, they decline to catch insects, sitting motionless all day long—pretty, if you like, but useless. The fact is, that all these creatures are nocturnal of habit. Very few men visit their orchid-houses at night, as I do constantly. They would see the frogs active enough then, creeping with wondrous dexterity among the leaves, and springing like a [Pg 15]green flash upon their prey. Naturally, therefore, they do not catch thrips or mealy-bug or aphis; these are too small game for the midnight sports-man. Wood-lice, centipedes, above all, cockroaches, those hideous and deadly foes of the orchid, are their victims. All who can keep them safe should have green frogs by the score in every house which they do not fumigate.

I have come to the orchids at last. It follows, indeed, almost of necessity that a man who has travelled much, an enthusiast in horticulture, should drift into that branch as years advance. Modesty would be out of place here. I have had successes, and if it please Heaven, I shall win more. But orchid culture is not to be dealt with at the end of an article.

III.

In the days of my apprenticeship I put up a big greenhouse: unable to manage plants in the open-air, I expected to succeed with them under unnatural conditions! These memories are strung together with the hope of encouraging a forlorn and desperate amateur here or there; and surely [Pg 16]that confession will cheer him. However deep his ignorance, it could not possibly be more finished than mine some dozen years ago; and yet I may say, *Je suis arrivé*! What that greenhouse cost, "chilled remembrance shudders" to recall; briefly, six times the amount, at least, which I should find ample now. And it was all wrong when done; not a trace of the original arrangement remains at this time, but there are inherent defects. Nothing throve, of course—except the insects. Mildew seized my roses as fast as I put them in; camellias dropped their buds with rigid punctuality; azaleas were devoured by thrips; "bugs," mealy and scaly, gathered to the feast; geraniums and pelargoniums grew like giants, but declined to flower. I consulted the local authority who was responsible for the well-being of a dozen gardens in the neighbourhood—an expert with a character to lose, from whom I bought largely. Said he, after a thorough inspection: "This concrete floor holds the water; you must have it swept carefully night and morning." That worthy man had a large business. His advice was sought by scores of neighbours like myself. And I tell the story as a warning; for he represents no small section of his class. My plants wanted not less but a great deal more water on that villainous concrete floor.

[Pg 17]Despairing of horticulture indoors as out, I sometimes thought of orchids. I had seen much of them in their native homes, both East and West—enough to understand that their growth is governed by strict law. Other plants—roses and so forth—are always playing tricks. They must have this and that treatment at certain times, the nature of which could not be precisely described, even if gardening books were written by men used to carry all the points of a subject in their minds, and to express exactly what they mean. Experience alone, of rather a dirty and uninteresting class, will give the skill necessary for success. And then they commit villanies of ingratitude beyond explanation. I knew that orchids must be quite different. Each class demands certain conditions as a preliminary: if none of them can be provided, it is a waste of money to buy plants. But when the needful conditions are present, and the poor things, thus relieved of a ceaseless preoccupation, can attend to business, it follows like a mathematical demonstration that if you treat them in such and such a way, such and such results will assuredly ensue. I was not aware then that many defy the most patient analysis of cause and effect. That knowledge is familiar now; but it does not touch the argument. Those cases [Pg 18]also are governed by rigid laws, which we do not yet understand.

Therefore I perceived or suspected, at an early date, that orchid culture is, as one may say, the natural province of an intelligent and enthusiastic amateur who has not the technical skill required for growing common plants. For it is brain-work—the other mechanical. But I shared the popular notion—which seems so very absurd now—that they are costly both to purchase and to keep: shared it so ingenuously that I never thought to ask myself how or why they could be more expensive, after the first outlay, than azaleas or gardenias. And meanwhile I was laboriously and impatiently gathering some comprehension of the ordinary plants. It was accident which broke the spell of ignorance. Visiting Stevens' Auction Rooms one day to buy bulbs, I saw a *Cattleya Mossiæ*, in bloom, which had not found a purchaser at the last orchid sale. A lucky impulse tempted me to ask the price. "Four shillings," said the invaluable Charles. I could not believe it—there must be a mistake: as if Charles ever made a mistake in his life! When he repeated the price, however, I seized that precious Cattleya, slapped down the money, and fled with it along King Street, fearing pursuit. Since no one followed, and Messrs. [Pg 19] Stevens did not write within the next few days reclaiming my treasure, I pondered the incident calmly. Perhaps they had been selling bankrupt stock, and perhaps they often do so. Presently I returned.

"Charles!" I said, "you sold me a *Cattleya Mossiæ* the other day."

Charles, in shirt-sleeves of course, was analyzing and summing up half a hundred loose sheets of figures, as calm and sure as a calculating machine. "I know I did, sir," he replied, cheerfully.

"It was rather dear, wasn't it?" I said.

"That's your business, sir," he laughed.

"Could I often get an established plant of *Cattleya Mossiæ* in flower for 4s.?" I asked.

"Give me the order, and I'll supply as many as you are likely to want within a month."

That was a revelation; and I tell the little story because I know it will be a revelation to many others. People hear of great sums paid for orchids, and they fancy that such represent only the extreme limits of an average. In fact, they have no relation whatsoever to the ordinary price. One of our largest general growers, who has but lately begun cultivating those plants, tells me that half-a-crown is the utmost he has paid for Cattleyas and Dendrobes, one shilling for Odonto [Pg 20]glots and Oncidiums. At these rates he has now a fine collection, many turning up among the lot for which he asks, and gets, as many pounds as the pence he gave. For such are imported, of course, and sold at auction as they arrive. This is not an article on orchids, but on "My Gardening," or I could tell some extraordinary tales. Briefly, I myself once bought a case two feet long, a foot wide, half-full of Odontoglossums for 8s. 6d. They were small bits, but perfect in condition. Of the fifty-three pots they made, not one, I think, has been lost. I sold the less valuable some years ago, when established and tested, at a fabulous profit. Another time I bought three "strings" of *O. Alexandræ*, the Pacho variety, which is finest, for 15s. They filled thirty-six pots, some three to a pot, for I could not make room for them all singly. Again—but this is enough. I only wish to demonstrate, for the service of very small amateurs like myself, that costliness at least is no obstacle if they have a fancy for this culture: unless, of course, they demand wonders and "specimens."

That *Cattleya Mossiæ*, was my first orchid, bought in 1884. It dwindled away, and many another followed it to limbo; but I knew enough, as has been said, to feel neither surprised nor angry. First of all, it is necessary to understand [Pg 21]the general conditions, and to secure them. Books give little help in this stage of education; they all lack detail in the preliminaries. I had not the good fortune to come across a friend or a gardener who grasped what was wrong until I found out for myself. For instance, no one told me that the concrete flooring of my house was a fatal error. When, a little disheartened, I made a new one, by glazing that *ruelle* mentioned in the preliminary survey of my garden, they allowed me to repeat it. Ingenious were my contrivances to keep the air moist, but none answered. It is not easy to find a material trim and clean which can be laid over concrete, but unless one can discover such, it is useless to grow orchids. I have no doubt that ninety-nine cases of failure in a hundred among amateurs are due to an unsuitable flooring. Glazed tiles, so common, are infinitely worst of all. May my experience profit others in like case!

Looking over the trade list of a man who manufactures orchid-pots one day, I observed, "Sea-sand for Garden Walks," and the preoccupation of years was dissipated. Sea-sand will hold water, yet will keep a firm, clean surface; it needs no rolling, does not show footprints nor muddy a visitor's boots. By next evening the floors were covered therewith six inches deep, and forthwith [Pg 22]my orchids began to flourish—not only to live. Long since, of course, I had provided a supply of water from the main to each house for "damping down." All round them now a leaden pipe was fixed, with pin-holes twelve inches apart, and a length of indiarubber hose at the end to fix upon the "stand-pipe." Attaching this, I turn the cock, and from each tiny hole spurts forth a jet, which in ten minutes will lay the whole floor under water, and convert the house into a shallow pond; but five minutes afterwards not a sign of the deluge is visible. Then I felt the joys of orchid culture. Much remained to learn—much still remains. We have some five thousand species in cultivation, of which an alarming number demand some difference of treatment if one would grow them to perfection. The amateur does not easily collect nor remember all this, and he is apt to be daunted if he inquire too deeply before "letting himself go." Such in especial I would encourage. Perfection is always a noble aim; but orchids do not exact it—far from that! The dear creatures will struggle to fulfil your hopes, to correct your errors, with pathetic patience. Give them but a chance, and they will await the progress of your education. That chance lies, as has been said, in the general conditions—the degree of moisture you [Pg 23]can keep in the air, the ventilation, and the light. These secured, you may turn up the books, consult the authorities, and gradually accumulate the knowledge which will enable you to satisfy the preferences of each class. So, in good time, you may enjoy such a thrill of pleasure as I felt the other day when a great pundit was good enough to pay me a call. He entered my tiny Odontoglossum house, looked round, looked round again, and turned to me. "Sir," he said, "we don't call this an amateur's collection!"

I have jotted down such hints of my experience as may be valuable to others, who, as Juvenal put it, own but a single lizard's run of earth. That space is enough to yield endless pleasure, amusement, and indeed profit, if a man cultivate it himself. Enthusiast as I am, I would not accept another foot of garden. [1]

FOOTNOTES:

[1] It is not inappropriate to record that when these articles were published in the *St. James' Gazette*, the editor received several communications warning him that his contributor was abusing his good faith—to put it in the mild French phrase. Happily, my friend was able to reply that he could personally vouch for the statements.

[Pg 24]

AN ORCHID SALE.

Shortly after noon on a sale day, the habitual customers of Messrs. Protheroe and Morris begin to assemble in Cheapside. On tables of roughest plank round the auction-rooms there, are neatly ranged the various lots; bulbs and sticks of every shape, big and little, withered or green, dull or shining, with a brown leaf here and there, or a mass of roots dry as last year's bracken. No promise do they suggest of the brilliant colours and strange forms buried in embryo within their uncouth bulk. On a cross table stand some dozens of "established" plants in pots and baskets, which the owners would like to part with. Their growths of this year are verdant, but the old bulbs look almost as sapless as those new arrivals. Very few are in flower just now—July and August are a time of pause betwixt the glories of the Spring and the milder effulgence of Autumn. Some great Dendrobes—*D. Dalhousianum*—are bursting into untimely bloom, betraying [Pg 25] to the initiated that their "establishment" is little more than a phrase. Those garlands of bud were conceived, so to speak, in Indian forests, have lain dormant through the long voyage, and began to show a few days since when restored to a congenial atmosphere. All our interest concentrates in the unlovely things along the wall.

The habitual attendants at an auction-room are always somewhat of a family party, but, as a rule, an ugly one. It is quite different with the regular group of orchid-buyers. No black sheep there. A dispute is the rarest of events, and when it happens everybody takes for granted that the cause is a misunderstanding. The professional growers are men of wealth, the amateurs men of standing at least. All know each other, and a cheerful familiarity rules. We have a duke in person frequently, who compares notes and asks a hint

27

from the authorities around; some clergymen; gentry of every rank; the recognized agents of great cultivators, and, of course, the representatives of the large trading firms. So narrow even yet is the circle of orchidaceans that almost all the faces at a sale are recognized, and if one wish to learn the names, somebody present can nearly always supply them. There is reason to hope that this will not be the case much [Pg 26]longer. As the mysteries and superstitions environing the orchid are dispersed, our small and select throng of buyers will be swamped, no doubt; and if a certain pleasing feature of the business be lost, all who love the flower and their fellow-men alike will cheerfully submit.

The talk is of orchids mostly, as these gentlemen stroll along the tables, lifting a root and scrutinizing it with practised glance that measures its vital strength in a second. But nurserymen take advantage of the gathering to show any curious or striking flower they chance to have at the moment. Mr. Bull's representative goes round, showing to one and another the contents of a little box—a lovely bloom of *Aristolochia elegans*, figured in dark red on white ground like a sublime cretonne—and a new variety of Impatiens; he distributes the latter presently, and gentlemen adorn their coats with the pale crimson flower.

Excitement does not often run so high as in the times, which most of those present can recall, when orchids common now were treasured by millionaires. Steam, and the commercial enterprise it fosters, have so multiplied our stocks, that shillings—or pence, often enough—represent the guineas of twenty years back. There are many here, scarcely yet grey, who could describe [Pg 27]the scene when *Masdevallia Tovarensis* first covered the stages of an auction-room. Its dainty white flowers had been known for several years. A resident in the German colony at Tovar, New Granada, sent one plant to a friend at Manchester, by whom it was divided. Each fragment brought a great sum, and the purchasers repeated this operation as fast as their morsels grew. Thus a conventional price was established—one guinea per leaf. Importers were few in those days, and the number of Tovars in South America bewildered them. At length Messrs. Sander got on the track, and commissioned Mr. Arnold to solve the problem. Arnold was a man of great energy and warm temper. Legend reports that he threw up the undertaking once because a gun offered him was second-hand; his prudence was vindicated afterwards by the misfortune of a *confrère*, poor Berggren, whose second-hand gun, presented by a Belgian employer, burst at a critical moment and crippled him for life. At the very moment of starting, Arnold had trouble with the railway officials. He was taking a quantity of Sphagnum moss in which to wrap the precious things, and they refused to let him carry it by passenger train. The station-master at Waterloo had never felt the atmosphere so warm, [Pg 28]they say. In brief, this was a man who stood no nonsense.

A young fellow-passenger showed much sympathy while the row went on, and Arnold learned with pleasure that he also was bound for Caraccas. This young man, whose name it is not worth while to cite, presented himself as agent for a manufacturer of Birmingham goods. There was no need for secrecy with a person of that sort. He questioned Arnold about orchids with a blank but engaging ignorance of the subject, and before the voyage was over he had learned all his friend's hopes and projects. But the deception could not be maintained at Caraccas. There Arnold discovered that the hardware agent was a collector and grower of orchids sufficiently well known. He said nothing, suffered his rival to start, overtook him at a village where the man was taking supper, marched in, barred the door, sat down opposite, put a revolver on the table, and invited him to draw. It should be a fair fight, said Arnold, but one of the pair must die. So convinced was the traitor of his earnestness—with good reason, too, as Arnold's acquaintances declare—that he slipped under the table, and discussed terms of abject surrender from that retreat. So, in due time, Messrs. Sander received more than forty thousand plants [Pg 29]of *Masdevallia Tovarensis*—sent them direct to the auction-room—and drove down the price in one month from a guinea a leaf to the fraction of a shilling.

Other great sales might be recalled, as that of *Phalænopsis Sanderiana* and *Vanda Sanderiana*, when a sum as yet unparalleled was taken in the room; *Cypripedium Spicerianum, Cyp. Curtisii, Lælia anceps alba*. Rarely now are we thrilled by sensations like these. But 1891 brought two of the old-fashioned sort, the reappearance of *Cattleya labiata autumnalis* and the public sale of *Dendrobium phalænopsis Schroderianum*. The former event deserves a special article, "The Lost

Orchid;" but the latter also was most interesting. Messrs. Sander are the heroes of both. *Dendrobium ph. Schroederianum* was not quite a novelty. The authorities of Kew obtained two plants from an island in Australasia a good many years ago. They presented a piece to Mr. Lee of Leatherhead, and another to Baron Schroeder; when Mr. Lee's grand collection was dispersed, the Baron bought his plant also, for £35, and thus possessed the only specimens in private hands. His name was given to the species.

Under these conditions, the man lucky and [Pg 30]enterprising enough to secure a few cases of the Dendrobium might look for a grand return. It seemed likely that New Guinea would prove to be its chief habitat, and thither Mr. Micholitz was despatched. He found it without difficulty, and collected a great number of plants. But then troubles began. The vessel which took them aboard caught fire in port, and poor Micholitz escaped with bare life. He telegraphed the disastrous news, "Ship burnt! What do?" "Go back," replied his employer. "Too late. Rainy season," was the answer. "Go back!" Mr. Sander repeated. Back he went.

This was in Dutch territory. "Well," writes Mr. Micholitz, "there is no doubt these are the meanest people on earth. On my telling them that it was very mean to demand anything from a shipwrecked man, they gave me thirty per cent. deduction on my passage"—201 dollars instead of 280 dollars. However, he reached New Guinea once more and tried fresh ground, having exhausted the former field. Again he found the Dendrobiums, of better quality and in greater number than before. But they were growing among bones and skeletons, in the graveyard of the natives. Those people lay their dead in a slight coffin, which they place upon the rocks just [Pg 31]above high tide, a situation which the Dendrobes love. Mr. Micholitz required all his tact and all his most attractive presents before he could persuade the Papuans to let him even approach. But brass wire proved irresistible. They not only suffered him to disturb the bones of their ancestors, but even helped him to stow the plunder. One condition they made: that a favourite idol should be packed therewith; this admitted, they performed a war dance round the cases, and assisted in transporting them. All went well this time, and in due course the tables were loaded with thousands of a plant which, before the consignment was announced, had been the special glory of a collection which is among the richest of the universe.

There were two memorable items in this sale: the idol aforesaid and a skull to which one of the Dendrobes had attached itself. Both were exhibited as trophies and curiosities, not to be disposed of; but by mistake, the idol was put up. It fetched only a trifle—quite as much as it was worth, however. But Hon. Walter de Rothschild fancied it for his museum, and on learning what had happened Mr. Sander begged the purchaser to name his own price. That individual refused.

It was a great day indeed. Very many of the [Pg 32]leading orchid-growers of the world were present, and almost all had their gardeners or agents there. Such success called rivals into the field, but New Guinea is a perilous land to explore. Only last week we heard that Mr. White, of Winchmore Hill, has perished in the search for *Dendrobium ph. Schroederianum*.

I mentioned the great sale of *Cyp. Curtisi* just now. An odd little story attaches to it. Mr. Curtis, now Director of the Botanic Gardens, Penang, sent this plant home from Sumatra when travelling for Messrs. Veitch, in 1882. The consignment was small, no more followed, and *Cyp. Curtisi* became a prize. Its habitat was unknown. Mr. Sander instructed his collector to look for it. Five years the search lasted—with many intermissions, of course, and many a success in discovering other fine things. But Mr. Ericksson despaired at last. In one of his expeditions to Sumatra he climbed a mountain—it has been observed before that one must not ask details of locality when collecting orchid legends. So well known is this mountain, however, that the Government, Dutch I presume, has built a shelter for travellers upon it. There Mr. Ericksson put up for the night. Several Europeans had inscribed their names upon the wall, with reflections [Pg 33]and sentiments, as is the wont of people who climb mountains. Among these, by the morning light, Mr. Ericksson perceived the sketch of a Cypripedium, as he lay upon his rugs. It represented a green flower, white tipped, veined and spotted with purple, purple of lip. "*Curtisi*, by Jove!" he cried, in his native Swedish, and jumped up. No doubt of it! Beneath the drawing ran: "C.C.'s contribution to the adornment of this house." Whipping out his pencil, Mr. Ericksson wrote: "Contribution accepted. Cypripedium collected!—C.E." But day by day he sought the plant in vain. His cases filled with other treasures. But for the hope that sketch conveyed, long since he would have left the spot. After all, Mr. Curtis might have chosen the flower by mere chance to decorate the wall. The natives did not know it. So orders were given to pack, and next day Mr. Ericksson would have withdrawn. On the very evening, however, one of his men brought in the flower. A curious story, if one think, but I am in a position to guarantee its truth.

Of another class, but not less renowned in its way, was the sale of March 11th last year. It had been heavily advertised. A leading continental importer announced the discovery of a new Odontoglossum. No less than six varieties [Pg 34]of type were employed to call public attention to its merits, and this was really no extravagant allowance under the circumstances alleged. It was a "grand new species," destined to be a "gem in the finest collections," a "favourite," the "most attractive of plants." Its flowers were wholly "tinged with a most delicate mauve, the base of the segment and the lip of a most charming violet"—in short, it was "the blue Odontoglossum" and well deserved the title *cœleste*. And the whole stock of two hundred plants would be offered to British enthusiasm. No wonder the crowd was thick at Messrs. Protheroe's room on that March morning. Few leading amateurs or growers who could not attend in person were unrepresented. At the psychological moment, when eagerness had reached the highest pitch, an orchid was brought in and set before them. Those experienced persons glanced at it and said, "Very nice, but haven't you an *Odontoglossum cœleste* to show?" The unhappy agent protested that this was the divine thing. No one would believe at first; the joke was too good—to put it in that mild form. When at length it became evident that this grand new species, heavenly gem, &c., was the charming but familiar *Odontoglossum ramossissimum*, such a tumult of laughter and indignation arose, that Messrs. [Pg 35] Protheroe quashed the sale. A few other instances of the kind might be given but none so grand.

The special interest of the sale to us lies in some novelties collected by Mr. Edward Wallace in parts unknown, and he is probably among us. Mr. Wallace has no adventures in particular to relate this time, but he tells, with due caution, where and how his treasures were gathered in South America. There is a land which those who have geographical knowledge sufficient may identify, surrounded by the territories of Peru, Ecuador, Colombia, Venezuela, and Brazil. It is traversed by some few Indian tribes, and no collector hitherto had penetrated it. Mr. Wallace followed the central line of mountains from Colombia for a hundred and fifty miles, passing a succession of rich valleys described as the loveliest ever seen by this veteran young traveller, such as would support myriads of cattle. League beyond league stretches the "Pajadena grass," pasturage unequalled; but "the wild herds that never knew a fold" are its only denizens. Here, on the mountain slopes, Mr. Wallace found *Bletia Sherrattiana*, the white form, very rare; another terrestrial orchid, unnamed and, as is thought, unknown, which sends up a branching spike two feet to three feet high, bearing ten to twelve flowers, of rich purple hue, in shape like a [Pg 36] Sobralia, three and four inches across; and yet another of the same family, growing on the rocks, and "looking like masses of snow on the hill-side." Such descriptions are thrilling, but these gentlemen receive them placidly; they would like to know, perhaps, what is the reserve price on such fine things, and what the chance of growing them to a satisfactory result. Dealers have a profound distrust of novelties, especially those of terrestrial genus; and their feeling is shared, for a like reason, by most who have large collections. Mr. Burbidge estimates roughly that we have fifteen hundred to two thousand species and varieties of orchid in cultivation; a startling figure, which almost justifies the belief of those who hold that no others worth growing will be found in countries already explored. But beyond question there are six times this number in existence, which collectors have not taken the trouble to gather. The chances, therefore, are against any new thing. Many species well known show slight differences of growth in different localities. Upon the whole, regular orchidaceans prefer that some one else should try experiments, and would rather pay a good price, when assured that it is worth their while, than a few shillings when the only certainty is trouble and the strong probability is failure. Mr. [Pg 37] Wallace has nothing more to tell of the undiscovered country. The Indians received him with composure, after he had struck up friendship with an old woman, and for the four days of his stay made themselves both useful and agreeable in their fashion.

The auctioneer has been chatting among his customers. He feels an interest in his wares, as who would not that dealt in objects of the extremest beauty and fascination? To him are consigned occasionally plants of unusual class, which the owner regards as unique, and expects to sell at the fanciest of prices. Unique indeed they must be which can pass unchallenged the ordeal of those keen and learned eyes. *Plumeria alba*, for instance, may be laid before them, and by no inexperienced horticulturist, with such a "reserve" as befits one of the most exquisite flowers known, and the only specimen in England. But a quiet smile goes round, and a gentleman present offers, in an audible whisper, to send in a dozen of that next week at a fraction of the price. So pleasant chat goes on, until, at the stroke of half-past twelve, the auctioneer mounts his rostrum. First to come before him are a hundred lots of *Odontoglossum crispum Alexandræ*, described as of "the very [Pg 38]best type, and in splendid condition." For the latter point everyone present is able to judge, and for the former all are willing to accept the statements of vendors. The glossy bulbs are clean as new pins, with the small "eye" just bursting among their roots; but nobody seems to want *Odontoglossum Alexandræ* in particular. One neat little bunch is sold for 11s., which will surely bear a wreath of white flowers, splashed with red brown, in the spring—perhaps two. And then bidding ceases. The auctioneer exclaims, "Does anybody want any *crispums*?" and instantly passes by the ninety-nine lots remaining.

It would mislead the unlearned public, and would not greatly interest them, to go through the catalogue of an orchid sale and quote the selling price of every lot. From week to week the value of these things fluctuates—that is, of course, of bulbs imported and unestablished. Various circumstances effect it, but especially the time of year. They sell best in spring, when they have months of light and sun before them, in which to recover from the effects of a long voyage and uncomfortable quarters. The buyer must make them grow strong before the dark days of an English winter are upon him; and every month that passes weakens his chance. In August it is [Pg 39]already late; in September, the periodical auctions ceased until lately. Some few consignments will be received, detained by accident, or forwarded by persons who do not understand the business.

That instance of *Odontoglossum Alexandræ* shows well enough the price of orchids this month, and the omission of all that followed illustrates it. The same lots would have been eagerly contested at twice the sum in April. But those who want that queenliest of flowers may get it for shillings at any time. The reputation of the importer, and his assurance that the plants belong to the very best type, give these more value than usual. He will try his luck once more perhaps this season; and then he will pot the bulbs unsold to offer them as "established" next year.

Oncidium luridum follows the Odontoglots, a broad-leaved, handsome orchid, which the untrained eye might think to have no pseudo-bulb at all. This species always commands a sale, if cheap, and ten shillings is a reasonable figure for a piece of common size. If all go well, it may throw out a branching spike six or seven feet long next summer, with — such a sight has been offered — several hundred blooms, yellow, brown and orange, *Oncidium juncifolium*, which comes next, is un [Pg 40]known to us, and probably to others; no offer is made for its reed-like growths described as "very free blooming all the year round, with small yellow flowers." *Epidendrum bicornutum*, on the other hand, is very well known and deeply admired, when seen; but this is an event too rare. The description of its exquisite white blossoms, crimson spotted on the lip, is still rather a legend than a matter of eye-witness. Somebody is reported to have grown it for some years "like a cabbage;" but his success was a mystery to himself. At Kew they find no trouble in certain parts of a certain house. Most of these, however, are fine growths, and the average price should be 12s. 6d. to 15s. Compare such figures with those that ruled when the popular impression of the cost of orchids was forming. I have none at hand which refer to the examples mentioned, but in the cases following, one may safely reckon shillings at the present day for pounds in 1846. That year, I perceive, such common species as *Barkeria spectabilis* fetched 5*l.* to 17*l.* each; *Epidendrum Stamfordianum*, five guineas; *Dendrobium formosum*, fifteen guineas; *Aerides maculosum*, *crispum* and *odoratum* 20*l.*, 21*l.*, and 16*l.*, respectively. No one who understands orchids will believe that the specimens which brought such monstrous prices [Pg 41]were superior in any respect to those we now receive, and he will be absolutely sure that they were landed in much worse condition. But the average cost of the most expensive at the present day might be 30s., and only a large piece would fetch that sum. It is astonishing to me that so few people grow orchids. Every modern book on gardening tells how five hundred varieties at least, the freest to flower and assuredly as beautiful as any, may be cultivated without heat for seven or eight months of the year. It is those "legends," I have spoken of which deter the public from entertaining the notion. An afternoon at an orchid sale would dispel them.

[Pg 42]

ORCHIDS.

There is no room to deal with this great subject historically, scientifically, or even practically, in the space of a chapter. I am an enthusiast, and I hold some strong views, but this is not the place to urge them. It is my purpose to ramble on, following thoughts as they arise, yet with a definite aim. The skilled reader will find nothing to criticize, I hope, and the indifferent, something to amuse.

Those amiable theorists who believe that the resources of Nature, if they be rightly searched, are able to supply every wholesome want the fancy of man conceives, have a striking instance in the case of orchids. At the beginning of this century, the science of floriculture, so far as it went, was at least as advanced as now. Under many disadvantages which we escape—the hot-air flue especially, and imperfect means of ventilation—our fore-fathers grew the plants known to them quite as well as we do. Many tricks have been discovered since, but for lasting success assuredly our systems are no improvement. Men interested in such [Pg 43]matters began to long for fresh fields, and they knew where to look. Linnæus had told them something of exotic orchids in 1763, though his knowledge was gained through dried specimens and drawings. One bulb, indeed—we spare the name—showed life on arrival, had been planted, and had flowered thirty years before, as Mr. Castle shows. Thus horticulturists became aware, just when the information was most welcome, that a large family of plants unknown awaited their attention; plants quite new, of strangest form, of mysterious habits, and beauty incomparable. Their notions were vague as yet, but the fascination of the subject grew from year to year. Whilst several hundred species were described in books, the number in cultivation, including all those gathered by Sir Joseph Banks, and our native kinds, was only fifty. Kew boasted no more than one hundred and eighteen in 1813; amateurs still watched in timid and breathless hope.

Gradually they came to see that the new field was open, and they entered with a rush. In 1830 a number of collections still famous in the legends of the mystery are found complete. At the Orchid Conference, Mr. O'Brien expressed a "fear that we could not now match some of the specimens mentioned at the exhibitions of the Horticultural [Pg 44] Society in Chiswick Gardens between 1835 and 1850;" and extracts which he gave from reports confirm this suspicion. The number of species cultivated at that time was comparatively small. People grew magnificent "specimens" in place of many handsome pots. We read of things amazing to the experience of forty years later. Among the contributions of Mrs. Lawrence, mother to our "chief," Sir Trevor, was an Aerides with thirty to forty flower spikes; a Cattleya with twenty spikes; an *Epidendrum bicornutum*, difficult to keep alive, much more to bloom, until the last few years, with "many spikes;" an Oncidium, "bearing a head of golden flowers four feet across." Giants dwelt in our greenhouses then.

So the want of enthusiasts was satisfied. In 1852 Mr. B.S. Williams could venture to publish "Orchids for the Million," a hand-book of world-wide fame under the title it presently assumed, "The Orchid Grower's Manual." An occupation or amusement the interest of which grows year by year had been discovered. All who took trouble to examine found proof visible that these masterworks of Nature could be transplanted and could be made to flourish in our dull climate with a regularity and a certainty unknown to them at home. The difficulties of their culture were found [Pg 45]to be a myth—we speak generally, and this point must be mentioned again. The "Million" did not yet heed Mr. Williams' invitation, but the Ten Thousand did, heartily.

I take it that orchids meet a craving of the cultured soul which began to be felt at the moment when kindly powers provided means to satisfy it. People of taste, unless I err, are tiring of those conventional forms in which beauty has been presented in all past generations. It may be an unhealthy sentiment, it may be absurd, but my experience is that it exists and must be taken into account. A picture, a statue, a piece of china, any work of art, is eternally the same, however charming. The most one can do is to set it in different positions, different lights. Théophile Gautier declared in a moment of frank impatience that if the Transfiguration hung in his study, he would assuredly find blemishes therein after awhile—quite fanciful and baseless, as he knew, but such, nevertheless, as would drive him to distraction presently. I entertain a notion, which may appear very odd to some, that Gautier's influence on the æsthetic class of men has been more vigorous than that of any other teacher; thousands who never read a line of his writing are unconsciously inspired by him. The feeling that gave birth to his protest [Pg 46]nearly two generations since is in the air now. Those who own a collection of art, those who have paid a great sum for pictures, will not allow it, naturally. As a rule, indeed, a man looks at his fine things no more than at his chairs and tables. But he who is best able to appreciate good work, and loves it best when he sees it, is the one who grows restless when it stands constantly before him.

"Oh, that those lips had language!" cried Cowper. "Oh, that those lovely figures would combine anew—change their light—do anything, anything!" cries the æsthete after awhile. "Oh, that the wind would rise upon that glorious sea; the summer green would fade to autumn yellow; that night would turn to day, clouds to sunshine, or sunshine to clouds." But the *littera scripta manet*—the stroke of the brush is everlasting. Apollo always bends the bow in marble. One may read a poem till it is known by heart, and in another second the familiar words strike fresh upon the ear. Painters lay a canvas aside, and presently come to it, as they say, with a new eye; but a purchaser once seized with this desperate malady has no such refuge. After putting his treasure away for years, at the first glance all his satiety returns. I myself have diagnosed a case where a fine drawing by Gerôme grew to be a veritable [Pg 47]incubus. It is understood that the market for pictures is falling yearly. I believe that the growth of this dislike to the eternal stillness of a painted scene is a chief cause of the disaster. It operates among the best class of patrons.

For such men orchids are a blessed relief. Fancy has not conceived such loveliness, complete all round, as theirs—form, colour, grace, distribution, detail, and broad effect. Somewhere, years ago—in Italy perhaps, but I think at the Taylor Institution, Oxford—I saw the drawings made by Rafaelle for Leo X. of furniture and decoration in his new palace; be it observed in parenthesis, that one who has not beheld the master's work in this utilitarian style of art has but a limited understanding of his supremacy. Among them were idealizations of flowers, beautiful and marvellous as fairyland, but compared with the glory divine that dwells in a garland of *Odontoglossum Alexandræ*, artificial, earthy. Illustrations of my meaning are needless to experts, and to others words convey no idea. But on the table before me now stands a wreath of *Oncidium crispum* which I cannot pass by. What colourist would dare to mingle these lustrous browns with pale gold, what master of form could shape the bold yet dainty waves and crisps and curls in its broad [Pg 48]petals, what human imagination could bend the graceful curve, arrange the clustering masses of its bloom? All beauty that the mind can hold is there—the quintessence of all charm and fancy. Were I acquainted with an atheist who, by possibility, had brain and feeling, I would set that spray before him and await reply. If Solomon in all his glory was not arrayed like a lily of the field, the angels of heaven have no vesture more ethereal than the flower of the orchid. Let us take breath.

Many persons indifferent to gardening—who are repelled, indeed, by its prosaic accompaniments, the dirt, the manure, the formality, the spade, the rake, and all that—love flowers nevertheless. For such these plants are more than a relief. Observe my Oncidium. It stands in a pot, but this is only for convenience—a receptacle filled with moss. The long stem feathered with great blossoms springs from a bare slab of wood. No mould nor peat surrounds it; there is absolutely nothing save the roots that twine round their support, and the wire that sustains it in the air. It asks no attention beyond its daily bath. From the day I tied it on that block last year—reft from home and all its pleasures, bought with paltry silver at Stevens' Auction Rooms—I have not touched it save to dip and to replace it on its hook. When the [Pg 49]flowers fade, thither it will return, and grow and grow, please Heaven, until next summer it rejoices me again; and so, year by year, till the wood rots. Then carefully I shall transfer it to a larger perch and resume. Probably I shall sever the bulbs without disturbing them, and in seasons following two spikes will push — then three, then a number, multiplying and multiplying when my remotest posterity is extinct. That is, so Nature orders it; whether my descendants will be careful to allow her fair play depends on circumstances over which I have not the least control.

For among their innumerable claims to a place apart among all things created, orchids may boast immortality. Said Sir Trevor Lawrence, in the speech which opened our famous Congress, 1885: "I do not see, in the case of most of them, the least reason why they should ever die. The parts of the orchideæ are annually reproduced in a great many instances, and there is really no reason they should not live for ever unless, as is generally the case with them in captivity, they be killed by errors in cultivation." Sir Trevor was addressing an assemblage of authorities — a parterre of kings in the empire of botany — or he might have enlarged upon this text.

The epiphytal orchid, to speak generally, and to [Pg 50]take the simple form, is one body with several limbs, crowned by one head. Its circulation pulsates through the whole, less and less vigorously, of course, in the parts that have flowered, as the growing head leaves them behind. At some age, no doubt, circulation fails altogether in those old limbs, but experience does not tell me distinctly as yet in how long time the worn-out bulbs of an Oncidium or a Cattleya, for example, would perish by natural death. One may cut them off when apparently lifeless, even beginning to rot, and under proper conditions — it may be a twelvemonth after — a tiny green shoot will push from some "eye," withered and invisible, that has slept for years, and begin existence on its own account. Thus, I am not old enough as an orchidacean to judge through how many seasons these plants will maintain a limb apparently superfluous. Their charming disposition is characterized above all things by caution and foresight. They keep as many strings to their bow, as many shots in their locker, as may be, and they keep them as long as possible. The tender young head may be nipped off by a thousand chances, but such mishaps only rouse the indomitable thing to replace it with two, or even more. Beings designed for immortality are hard to kill.

Among the gentle forms of intellectual excitement I know not one to compare with the joy of restoring a neglected orchid to health. One may buy such for coppers—rare species, too—of a size and a "potentiality" of display which the dealers would estimate at as many pounds were they in good condition on their shelves. I am avoiding names and details, but it will be allowed me to say, in brief, that I myself have bought more than twenty pots for five shillings at the auction-rooms, not twice nor thrice either. One half of them were sick beyond recovery, some few had been injured by accident, but by far the greater part were victims of ignorance and ill-treatment which might still be redressed. Orchids tell their own tale, whether of happiness or misery, in characters beyond dispute. Mr. O'Brien alleged, indeed, before the grave and experienced signors gathered in conference, that "like the domestic animals, they soon find out when they are in hands that love them. With such a guardian they seem to be happy, and to thrive, and to establish an understanding, indicating to him their wants in many important matters as plainly as though they could speak." And the laugh that followed this statement was not derisive. He who glances at the endless tricks, methods, and contrivances devised by one or other species to serve its turn may well come to fancy that orchids are reasoning things.

At least, many keep the record of their history in form unmistakable. Here is a Cattleya which I purchased last autumn, suspecting it to be rare and valuable, though nameless; I paid rather less than one shilling. The poor thing tells me that some cruel person bought it five years ago—an imported piece, with two pseudo-bulbs. They still remain, towering like columns of old-world glory above an area of shapeless ruin. To speak in mere prose—though really the conceit is not extravagant—these fine bulbs, grown in their native land, of course, measure eight inches high by three-quarters of an inch diameter. In the first season, that *malheureux* reduced their progeny to a stature of three and a half inches by the foot-rule; next season, to two inches; the third, to an inch and a half. By this time the patient creature had convinced itself that there was something radically wrong in the circumstances attending its normal head, and tried a fresh departure from the stock—a "back growth," as we call it, after the fashion I have described. In the third year then, there were two heads. In the fourth year, the chief of them had dwindled to less than one inch [Pg 53]and the thickness of a straw, while the second struggled into growth with pain and difficulty, reached the size of a grain of wheat, and gave it up. Needless to say that the wicked and unfortunate proprietor had not seen trace of a bloom. Then at length, after five years' torment, he set it free, and I took charge of the wretched sufferer. Forthwith he began to show his gratitude, and at this moment—the summer but half through—his leading head has regained all the strength lost in three years, while the back growth, which seemed dead, outtops the best bulb my predecessor could produce.

And I have perhaps a hundred in like case, cripples regaining activity, victims rescued on their death-bed. If there be a placid joy in life superior to mine, as I stroll through my houses of a morning, much experience of the world in many lands and many circumstances has not revealed it to me. And any of my readers can attain it, for—in no conventional sense—I am my own gardener; that is to say, no male being ever touches an orchid of mine.

One could hardly cite a stronger argument to demolish the superstitions that still hang around this culture. If a busy man, journalist, essayist, novelist, and miscellaneous *littérateur*, who lives by [Pg 54]his pen, can keep many hundreds of orchids in such health that he is proud to show them to experts—with no help whatsoever beyond, in emergency, that which ladies of his household, or a woman-servant give—if he can do this, assuredly the pursuit demands little trouble and little expense. I am not to lay down principles of cultivation here, but this must be said: orchids are indifferent to detail. There lies a secret. Secure the general conditions necessary for their well-doing, and they will gratefully relieve you of further anxiety; neglect those general conditions, and no care will reconcile them. The gentleman who reduced my Cattleya to such straits gave himself vast pains, it is likely, consulted no end of books, did all they recommend; and now declares that orchids are unaccountable. It is just the reverse. No living things follow with such obstinate obedience a few most simple laws; no machine produces its result more certainly, if one comply with the rules of its being.

This is shown emphatically by those cases which we do not clearly understand; I take for example the strangest, as is fitting. Some irreverent zealots have hailed the Phalœnopsis as Queen of Flowers, dethroning our venerable rose. I have not to consider the question of allegiance, but decidedly [Pg 55]this is, upon the whole, the most interesting of all orchids in the cultivator's point of view. For there are some genera and many species that refuse his attentions more or less stubbornly—in fact, we do not yet know how to woo them. But the Phalœnopsis is not among them. It gives no trouble in the great majority of cases. For myself, I find it grow with the calm complacency of the cabbage. Yet we are all aware that our success is accidental, in a measure. The general conditions which it demands are fulfilled, commonly, in any stove where East Indian plants flourish; but from time to time we receive a vigorous hint that particular conditions, not always forthcoming, are exacted by Phalœnopsis. Many legends on this theme are current; I may cite two, notorious and easily verified. The authorities at Kew determined to build a special house for the genus, provided with every comfort which experience or scientific knowledge could suggest. But when it was opened, six or eight years ago, not a Phalœnopsis of all the many varieties would grow in it; after vain efforts, Mr. Thiselton Dyer was obliged to seek another use for the building, which is now employed to show plants in flower. Sir Trevor Lawrence tells how he laid out six hundred pounds [Pg 56]for the same object with the same result. And yet one may safely reckon that this orchid does admirably in nine well-managed stoves out of ten, and fairly in nineteen out of twenty. Nevertheless, it is a maxim with growers that Phalœnopsis should never be transferred from a situation where they are doing well. Their hooks are sacred as that on which Horace suspended his lyre. Nor could a reasonable man think this fancy extravagant, seeing the evidence beyond dispute which warns us that their health is governed by circumstances more delicate than we can analyze at present.

It would be wrong to leave the impression that orchid culture is actually as facile as market gardening, but we may say that the eccentricities of Phalœnopsis and the rest have no more practical importance for the class I would persuade than have the terrors of the deep for a Thames water-man. How many thousand householders about this city have a "bit of glass" devoted to geraniums and fuchsias and the like! They started with more ambitious views, but successive disappointments have taught modesty, if not despair. The poor man now contents himself with anything that will keep tolerably green and show some spindling flower. The fact is, that hardy [Pg 57]plants under glass demand skilful treatment—all their surroundings are unnatural, and with insect pest on one hand, mildew on the other, an amateur stands betwixt the devil and the deep sea. Under those circumstances common plants become really capricious—that is, being ruled by no principles easy to grasp and immutable in operation, their discomfort shows itself in perplexing forms. But such species of orchids as a poor man would think of growing are incapable of pranks. For one shilling he can buy a manual which will teach him what these species are, and most of the things necessary for him to understand besides. An expenditure of five pounds will set him up for life and beyond—since orchids are immortal. Nothing else is needed save intelligence.

Not even heat, since his collection will be "cool" naturally; if frost be excluded, that is enough. I should not have ventured to say this some few years ago—before, in fact, I had visited St. Albans. But in the cool house of that palace of enchantment with which Mr. Sander has adorned the antique borough, before the heating arrangements were quite complete though the shelves were occupied, often the glass would fall very low into the thirties. I could never learn distinctly that mischief followed, though Mr. [Pg 58] Godseff did not like it at all. One who beheld the sight when those fields of Odontoglossum burst into bloom might well entertain a doubt whether improvement was possible. There is nothing to approach it in this lower world. I cannot forbear to indicate one picture in the grand gallery. Fancy a corridor four hundred feet long, six wide, roofed with square baskets hanging from the glass as close as they will fit. Suspend to each of these—how many hundreds or thousands has never been computed—one or more garlands of snowy flowers, a thicket overhead such as one might behold in a tropic forest, with myriads of white butterflies clustering amongst the vines. But imagination cannot bear mortal man thus far. "Upon the banks of Paradise" those "twa clerks" may have seen the like; yet, had they done so their hats would have been adorned not with "the birk," but with plumes of *Odontoglossum citrosmum*.

I have but another word to say. If any of the class to whom I appeal incline to let "I dare not wait upon I would," hear the experience of a bold enthusiast, as recounted by Mr. Castle in his small brochure, "Orchids." This gentleman had a fern-case outside his sitting-room window, six feet long by three wide. He ran pipes through it, warmed [Pg 59]presumably by gas. More ambitious than I venture to recommend, "in this miniature structure," says Mr. Castle, "with liberal supplies of water, the owner succeeded in growing, in a smoky district of London"—I will not quote the amazing list of fine things, but it numbers twenty-five species, all the most delicate and beautiful of the stove kinds. If so much could be done under such circumstances, what may rightly be called difficult in the cultivation of orchids?

[Pg 60]

COOL ORCHIDS.

This is a subject which would interest every cultured reader, I believe, every householder at least, if he could be brought to understand that it lies well within the range of his practical concerns. But the public has still to be persuaded. It seems strange to the expert that delusions should prevail when orchids are so common and so much talked of; but I know by experience that the majority of people, even among those who love their garden, regard them as fantastic and mysterious creations, designed, to all seeming, for the greater glory of pedants and millionaires. I try to do my little part, as occasion serves, in correcting this popular error, and spreading a knowledge of the facts. It is no less than a duty. If every human being should do what he can to promote the general happiness, it would be downright wicked to leave one's fellow-men under the influence of hallucinations that debar them from the most charming of quiet pleasures. I suspect also that the misapprehension of the public is largely due to the conduct of [Pg 61]experts in the past. It was a rule with growers formerly, avowed among themselves, to keep their little secrets. When Mr. B.S. Williams published the first edition of his excellent book forty years ago, he fluttered his colleagues sadly. The plain truth is that no class of plant can be cultivated so easily, as none are so certain to repay the trouble, as the Cool Orchids.

Nearly all the genera of this enormous family have species which grow in a temperate climate, if not in the temperate zone. At this moment, in fact, I recall but two exceptions, Vanda and Phalœnopsis. Many more there are, of course—half a dozen have occurred to me while I wrote the last six words—but in the small space at command I must cling to generalities. We have at least a hundred genera which will flourish anywhere if the frost be excluded; and as for species, a list of two thousand would not exhaust them probably. But a reasonable man may content himself with the great classes of Odontoglossum, Oncidium, Cypripedium, and Lycaste; among the varieties of these, which no one has ventured to calculate perhaps, he may spend a happy existence. They have every charm—foliage always green, a graceful habit, flowers that rank among the master works of Nature. The poor man who succeeds [Pg 62]with them in his modest "bit of glass" has no cause to envy Dives his flaunting Cattleyas and "fox-brush" Aerides. I should like to publish it in capitals—that nine in ten of those suburban householders who read this book may grow the loveliest of orchids if they can find courage to try.

Odontoglossums stand first, of course—I know not where to begin the list of their supreme merits. It will seem perhaps a striking advantage to many that they burst into flower at any time, as they chance to ripen. I think that the very perfection of culture is discounted somewhat in this instance. The gardener who keeps his plants at the *ne plus ultra* stage brings them all into bloom within the space of a few weeks. Thus in the great collections there is such a show during April, May, and June as the Gardens of Paradise could not excel, and hardly a spike in the cool houses for the rest of the year. At a large establishment this signifies nothing; when the Odontoglossums go off other things "come on" with equal regularity. But the amateur, with his limited assortment, misses every bloom. He has no need for anxiety with this genus. It is their instinct to flower in spring, of course, but they are not pedantic about it in the least. Some tiny detail overlooked here and there, [Pg 63]absolutely unimportant to health, will retard florescence. It might very well happen that the owner of a dozen pots had one blooming every month successively. And that would mean two spikes open, for, with care, most Odontoglossums last above four weeks.

Another virtue, shared by others of the cool class in some degree, is their habit of growing in winter. They take no "rest;" all the year round their young bulbs are swelling, graceful foliage lengthening, roots pushing, until the spike demands a concentration of all their energy. But winter is the most important time. I think any man will see the peculiar blessing of this arrangement. It gives interest to the long dull days, when other plant life is at a standstill. It furnishes material for cheering meditations on a Sunday morning—is that a trifle? And at this season the pursuit is joy unmixed. We feel no anxious questionings, as we go about our daily business, whether the *placens uxor* forgot to remind Mary, when she went out, to pull the blinds down; whether Mary followed the instructions if given; whether those confounded patent ventilators have snapped to again. Green fly does not harass us. One syringing a day, and one watering per week suffice. Truly these are not grave things, but the issue at [Pg 64]stake is precious: we enjoy the boon of relief proportionately.

Very few of those who grow Odontoglossums know much about the "Trade," or care, seemingly. It is a curious subject, however. The genus is American exclusively. It ranges over the continent from the northern frontier of Mexico to the southern frontier of Peru, excepting, to speak roughly, the empire of Brazil. This limitation is odd. It cannot be due to temperature simply, for, upon the one hand, we receive Sophronitis, a very cool genus, from Brazil, and several of the coolest Cattleyas; upon the other, *Odontoglossum Roezlii*, a very hot species, and *O. vexillarium*, most decidedly warm, flourish up to the boundary. Why these should not step across, even if their mountain sisters refuse companionship with the Sophronitis, is a puzzle. Elsewhere, however, they abound. Collectors distinctly foresee the time when all the districts they have "worked" up to this will be exhausted. But South America contains a prodigious number of square miles, and a day's march from the track carries one into *terra incognita*. Still, the end will come. The English demand has stripped whole provinces, and now all the civilized world is entering into competition. We are sadly assured that Odontoglossums carried off will not be replaced [Pg 65]for centuries. Most other genera of orchid propagate so freely that wholesale depredations are made good in very few years. For reasons beyond our comprehension as yet, the Odontoglossum stands in different case. No one in England has raised a plant from seed—that we may venture to say definitely. Mr. Cookson and Mr. Veitch, perhaps others also, have obtained living germs, but they died incontinently. Frenchmen, aided by the climate, have been rather more successful. MM. Bleu and Moreau have both flowered seedling Odontoglots. M. Jacob, who takes charge of M. Edmund de Rothschild's orchids at Armainvilliers, has a considerable number of young plants. The reluctance of Odontoglots to propagate is regarded as strange; it supplies a constant theme for discussion among orchidologists. But I think that if we look more closely it appears consistent with other facts known. For among importations of every genus but this—and Cypripedium—a plant bearing its seed-capsules is frequently discovered; but I cannot hear of such an incident in the case of Odontoglossums. They have been arriving in scores of thousands, year by year, for half a century almost, and scarcely anyone recollects observing a seed-capsule. This shows how rarely they fertilize in their native home. When [Pg 66]that event happens, the Odontoglossum is yet more prolific than most, and the germs, of course, are not so delicate under their natural conditions. But the moral to be drawn is that a country once stripped will not be reclothed.

I interpolate here a profound observation of Mr. Roezl. That wonderful man remarked that Odontoglossums grow upon branches thirty feet above the ground. It is rare to find them at thirty-five feet, rarer at twenty-five; at greater and less heights they do not exist. Here, doubtless, we have the secret of their reluctance to fertilize; but I will offer no comments, because the more one reflects the more puzzling it becomes. Evidently the seed must be carried above and must fall below that limit, under circumstances which, to our apprehension, seem just as favourable as those at the altitude of thirty feet. But they do not germinate. Upon the other hand, Odontoglossums show no such daintiness of growth in our houses. They flourish at any height, if the general conditions be suitable. Mr.

Roezl discovered a secret nevertheless, and in good time we shall learn further.

To the Royal Horticultural Society of England belongs the honour of first importing orchids methodically and scientifically. Messrs. Weir and [Pg 67] Fortune, I believe, were their earliest employés. Another was Theodor Hartweg, who discovered *Odontoglossum crispum Alexandræ* in 1842; but he sent home only dried specimens. From these Lindley described and classed the plant, aided by the sketch of a Spanish or Peruvian artist, Tagala. A very curious mistake Lindley fell into on either point. The scientific error does not concern us, but he represented the colouring of the flower as yellow with a purple centre. So Tagala painted it, and his drawing survives. It is an odd little story. He certainly had Hartweg's bloom before him, and that certainly was white. But then again yellow Alexandræs have been found since that day. To the Horticultural Society we are indebted, not alone for the discovery of this wonder, but also for its introduction. John Weir was travelling for them when he sent living specimens in 1862. It is not surprising that botanists thought it new after what has been said. As such Mr. Bateman named it after the young Princess of Wales—a choice most appropriate in every way.

Odontoglossum Crispum Alexandrae.
Flower reduced to One Fourth
Flower Stem to One Sixth

Then a few wealthy amateurs took up the business of importation, such as the Duke of Devonshire. But "the Trade" came to see presently that there was money in this new fashion, and imported so vigorously that the [Pg 68]Society found its exertions needless. Messrs. Rollisson of Tooting, Messrs. Veitch of Chelsea, and Messrs. Low of Clapton distinguished themselves from the outset. Of these three firms one is extinct; the second has taken up, and made its own, the fascinating study of hybridization among orchids; the third still perseveres. Twenty years ago, nearly all the great nurserymen in London used to send out their travellers; but they have mostly dropped the practice. Correspondents forward a shipment from time to time. The expenses of the collector are heavy, even if he draw no more than his due—and the temptation to make up a fancy bill cannot be resisted by some weak mortals. Then, grave losses are always probable—in the case of South American importations, certain. It has happened not once but a hundred times that the toil of months, the dangers, the sufferings, and the hard money expended go to absolute waste. Twenty or thirty thousand plants or more an honest man collects, brings down from the mountains or the forests, packs carefully, and ships. The freight alone may reach from three to eight hundred pounds—I have personally known instances when it exceeded five hundred. The cases arrive in England—and not a living thing therein! [Pg 69] A steamship company may reduce its charge under such circumstances, but again and again it will happen that the speculator stands out of a thousand pounds clean when his boxes are opened. He may hope to recover it on the next cargo, but that is still a question of luck. No wonder that men whose business is not confined to orchids withdrew from the risks of importation, returning to roses and lilies and daffodowndillies with a new enthusiasm.

There is another point also, which has varying force with different characters. The loss of life among those men who "go out collecting" has been greater proportionately, than in any class of which I have heard. In former times, at least, they were chosen haphazard, among intelligent and trustworthy employés of the firm. Trustworthiness was a grand point, for reasons hinted. The honest youth, not very strong perhaps in an English climate, went bravely forth into the unhealthiest parts of unhealthy lands, where food is very scarce, and very, very rough; where he was wet through day after day, for weeks at a time; where "the fever," of varied sort, comes as regularly as Sunday; where from month to month he found no one with whom to exchange a word. I could make out a startling list of the martyrs of orchidology. Among Mr. Sander's collectors alone, Falkenberg [Pg 70]perished at Panama, Klaboch in Mexico, Endres at Rio Hacha, Wallis in Ecuador, Schroeder in Sierra Leone, Arnold on the Orinoco, Digance in Brazil, Brown in Madagascar. Sir Trevor Lawrence mentions a case where the zealous explorer "waded for a fortnight up to his middle in mud," searching for a plant he had heard of. I have not identified this instance of devotion, but we know of rarities which would demand perseverance and sufferings almost equal to secure them. If employers could find the heart to tempt a fellow-creature into such risks, the chances are that it would prove bad business. For to discover a new or valuable orchid is only the first step in a commercial enterprise. It remains to secure the "article," to bring it safely into a realm that may be called civilized, to pack it and superintend its transport through the sweltering lowland to a shipping place. If the collector sicken after finding his prize, these cares are neglected more or less; if he die, all comes to a full stop. Thus it happens that the importing business has been given up by one firm after another.

Odontoglossums, as I said, belong to America—to the mountainous parts of the continent in general. Though it would be wildly rash to pronounce which is the loveliest of orchids, no man [Pg 71]with eyes would dispute that *O. crispum Alexandræ* is the queen of this genus. She has her home in the States of Colombia, and those who seek her make Bogota their headquarters. If the collector wants the broad-petalled variety, he goes about ten days to the southward before commencing operations; if the narrow-petalled, about two days to the north—on mule-back of course. His first care on arrival in the neighbourhood—which is unexplored ground, if such he can discover—is to hire a wood; that is, a track of mountain clothed more or less with timber. I have tried to procure one of these "leases," which must be odd documents; but orchid-farming is a close and secret business. The arrangement concluded in legal form, he hires natives, twenty or fifty or a hundred, as circumstances advise, and sends them to cut down trees, building meantime a wooden stage of sufficient length to bear the plunder expected. This is used for cleaning and drying the plants brought in. Afterwards, if he be prudent, he follows his lumber-men, to see that their indolence does not shirk the big trunks—which give extra trouble naturally, though they yield the best and largest return. It is a terribly wasteful process. If we estimate that a good tree has been felled for every three scraps of Odontoglossum which are now established in [Pg 72] Europe, that will be no exaggeration. And for many years past they have been arriving by hundreds of thousands annually! But there is no alternative. An European cannot explore that green wilderness overhead; if he could, his accumulations would be so slow and costly as to raise the proceeds to an impossible figure. The natives will not climb, and they would tear the plants to bits. Timber has no value in those parts as yet, but the day approaches when Government must interfere. The average yield of *Odontoglossum crispum* per tree is certainly not more than five large and small together. Once upon a time Mr. Kerbach recovered fifty-three at one felling, and the incident has grown into a legend; two or three is the usual number. Upon the other hand, fifty or sixty of *O. gloriosum*, comparatively worthless, are often secured. The cutters receive a fixed price of sixpence for each orchid, without reference to species or quality.

When his concession is exhausted, the traveller overhauls the produce carefully, throwing away those damaged pieces which would ferment in the long, hot journey home, and spoil the others. When all are clean and dry, he fixes them with copper wire on sticks, which are nailed across boxes for transport. Long experience has laid [Pg 73]down rules for each detail of this process. The sticks, for example, are one inch in diameter, fitting into boxes two feet three inches wide, two feet deep, neither more nor less. Then the long file of mules sets out for Bogota, perhaps ten days' march, each animal carrying two boxes—a burden ridiculously light, but on such tracks it is dimension which has to be considered. On arrival at Bogota, the cases are unpacked and examined for the last time, re-stowed, and consigned to the muleteers again. In six days they reach Honda, on the Magdalena River, where, until lately, they were embarked on rafts for a voyage of fourteen days to Savanilla. At the present time, an American company has established a service of flat-bottomed steamers which cover the distance in seven days, thus reducing the risks of the journey by one-half. But they are still terrible. Not a breath of wind stirs the air at that season, for the collector cannot choose his time. The boxes are piled on deck; even the pitiless sunshine is not so deadly as the stewing heat below. He has a store of blankets to cover them, on which he lays a thatch of palm-leaves, and all day long he souses the pile with water; but too well the poor fellow knows that mischief is busy down below. Another anxiety possesses him too. It may very well be that on [Pg 74]arrival at Savanilla he has to wait days in that sweltering atmosphere for the Royal Mail steamer. And when it comes in, his troubles do not cease, for the stowage of the precious cargo is vastly important. On deck it will almost certainly be injured by salt water. In the hold it will ferment. Amidships it is apt to be baked by the engine fire. Whilst writing I learn that Mr. Sander has lost two hundred and sixty-seven cases by this latter mishap, as is supposed. So utterly hopeless is their condition, that he will not go to the expense of overhauling them; they lie at Southampton, and to anybody who will take them away all parties concerned will be grateful. The expense of making this shipment a reader may judge from the hints given. The Royal Mail Company's charge for freight from Manzanilla is 750*l*. I could give an incident of the same class yet more startling with reference to Phalœnopsis. It is proper to add that the most enterprising of Assurance Companies do not yet see their way to accept any kind of risks in the orchid trade; importers must bear all the burden. To me it seems surprising that the plants can be sold so cheap, all things considered. Many persons think and hope that prices will fall, and that may probably happen with regard to some genera. But the shrewdest of those very shrewd [Pg 75]men who conduct the business all look for a rise.

Od. Harryanum always reminds me—in such an odd association of ideas as everyone has experienced—of a thunderstorm. The contrast of its intense brown blotches with the azure throat and the broad, snowy lip, affect me somehow with admiring oppression.

Very absurd; but *on est fait comme ça*, as Nana excused herself. To call this most striking flower "Harryanum" is grotesque. The public is not interested in those circumstances which give the name significance for a few, and if there be any flower which demands an expressive title, it is this, in my judgment. Possibly it was some Indian report which had slipped his recollection that led Roezl to predict the discovery of a new Odontoglot, unlike any other, in the very district where *Od. Harryanum* was found after his death, though the story is quoted as an example of that instinct which guides the heaven-born collector. The first plants came unannounced in a small box sent by Señor Pantocha, of Colombia, to Messrs. Horsman in 1885, and they were flowered next year by Messrs. Veitch. The dullest who sees it can now imagine the excitement when this marvel was displayed, coming from an unknown habitat. Roezl's [Pg 76]prediction occurred to many of his acquaintance, I have heard; but Mr. Sander had a living faith in his old friend's sagacity. Forthwith he despatched a collector to the spot which Roezl had named—but not visited—and found the treasure. The legends of orchidology will be gathered one day, perhaps; and if the editor be competent, his volume should be almost as interesting to the public as to the cognoscenti.

I have been speaking hitherto of Colombian Odontoglossums, which are reckoned among the hardiest of their class. Along with them, in the same temperature, grow the cool Masdevallias, which probably are the most difficult of all to transport. There was once a grand consignment of *Masdevallia Schlimii*, which Mr. Roezl despatched on his own account. It contained twenty-seven thousand plants of this species, representing at that time a fortune. Mr. Roezl was the luckiest and most experienced of collectors, and he took special pains with this unique shipment. Among twenty-seven thousand two bits survived when the cases were opened; the agent hurried them off to Stevens's auction-rooms, and sold them forthwith at forty guineas each. But I must stick to Odontoglossums. Speculative as is the business of importing the northern species, to gather those of [Pg 77] Peru and Ecuador is almost desperate. The roads of Colombia are good, the population civilized, conveniences abound, if we compare that region with the orchid-bearing territories of the south. There is a fortune to be secured by anyone who will bring to market a lot of *O. nœveum* in fair condition. Its habitat is perfectly well known. I am not aware that it has a delicate constitution; but no collector is so rash or so enthusiastic as to try that adventure again, now that its perils are understood; and no employer is so reckless as to urge him. The true variety of *O. Hallii* stands in much the same case. To obtain it the explorer must march in the bed of a torrent and on the face of a precipice alternately for an uncertain period of time, with a river to cross about every day. And he has to bring back his loaded mules, or Indians, over the same pathless waste. The Roraima Mountain begins to be regarded as quite easy travel for the orchid-hunter nowadays. If I mention that the canoe-work on this route demands thirty-two portages, thirty-two loadings and unloadings of the cargo, the reader can judge what a "difficult road" must be. Ascending the Roraima, Mr. Dressel, collecting for Mr. Sander, lost his herbarium in the Essequibo River. Savants alone are able to estimate the awful nature of the crisis [Pg 78]when a comrade looses his grip of that treasure. For them it is needless to add that everything else went to the bottom. [2]

One is tempted to linger among the Odontoglots, though time is pressing. In no class of orchids are natural hybrids so mysterious and frequent. Sometimes one can detect the parentage; in such cases, doubtless, the crossing occurred but a few generations back: as a rule, however, such plants are the result of breeding in and in from age to age, causing all manner of delightful complications. How many can trace the lineage of Mr. Bull's *Od. delectabile*—ivory white, tinged with rose, strikingly blotched with red and showing a golden labellum? or Mr. Sander's *Od. Alberti-Edwardi*, which has a broad soft margin of gold about its stately petals? Another is rosy white, closely splashed with pale purple, and dotted round the edge with spots of the same tint so thickly placed that they resemble a fringe. Such marvels turn up in an importation without the slightest warning—no peculiarity betrays them until the flowers open; when the lucky purchaser discovers that a plant for which he gave perhaps a shilling is worth an indefinite number of guineas.

Lycaste also is a genus peculiar to America, [Pg 79]such a favourite among those who know its merits that the species *L. Skinneri* is called the "Drawing-Room Flower." Professor Reichenbach observes in his superb volume that many people utterly ignorant of orchids grow this plant in their miscellaneous collection. I speak of it without prejudice, for to my mind the bloom is stiff, heavy, and poor in colour. But there are tremendous exceptions. In the first place, *Lycaste Skinneri alba*, the pure white variety, beggars all description. Its great flower seems to be sculptured in the snowiest of transparent marble. That stolid pretentious air which offends one—offends me, at least—in the coloured examples, becomes virginal dignity in this case. Then, of the normal type there are more than a hundred variations recognized, some with lips as deep in tone, and as smooth in texture, as velvet, of all shades from maroon to brightest crimson. It will be understood that I allude to the common forms in depreciating this species. How vast is the difference between them, their commercial value shows. Plants of the same size and the same species range from 3*s*. 6*d*. to 35 guineas, or more indefinitely.

Lycastes are found in the woods, of Guatemala especially, and I have heard no such adventures [Pg 80]in the gathering of them as attend Odontoglossums. Easily obtained, easily transported, and remarkably easy to grow, of course they are cheap. A man must really "give his mind to it" to kill a Lycaste. This counts for much, no doubt, in the popularity of the genus, but it has plenty of other virtues. *L. Skinneri* opens in the depth of winter, and all the rest, I think, in the dull months. Then, they are profuse of bloom, throwing up half a dozen spikes, or, in some species, a dozen, from a single bulb, and the flowers last a prodigious time. Their extraordinary thickness in every part enables them to withstand bad air and changes of temperature, so that ladies keep them on a drawing-room table, night and day, for months, without change perceptible. Mr. Williams names an instance where a *L. Skinneri*, bought in full bloom on February 2, was kept in a sitting-room till May 18, when the purchaser took it back, still handsome. I have heard cases more surprising. Of species somewhat less common there is *L. aromatica*, a little gem, which throws up an indefinite number of short spikes, each crowned with a greenish yellow triangular sort of cup, deliciously scented. I am acquainted with no flower that excites such enthusiasm among ladies who fancy Messrs. Liberty's style of toilette; sad ex [Pg 81]perience tells me that ten commandments or twenty will not restrain them from appropriating it. *L. cruenta* is almost as tempting. As for *L. leucanthe,* an exquisite combination of pale green and snow white, it ranks with *L. Skinneri alba* as a thing too beautiful for words. This species has not been long introduced, and at the moment it is dear proportionately. There is yet another virtue of the Lycaste which appeals to the expert. It lends itself readily to hybridization. This most fascinating pursuit attracts few amateurs as yet, and the professionals have little time or inclination for experiments. They naturally prefer to make such crosses as are almost certain to pay. Thus it comes about that the hybridization of Lycastes has been attempted but recently, and none of the seedlings, so far as I can learn, have flowered. They have been obtained, however, in abundance, not only from direct crossing, but also from alliance with Zygopetalum, Anguloa, and Maxillaria.

The genus Cypripedium, Lady's Slipper, is perhaps more widely scattered over the globe than any other class of plant; I, at least, am acquainted with none that approaches it. From China to Peru—nay, beyond, from Archangel to Torres Straits,—but it is wise to avoid these semi-poetic [Pg 82]descriptions. In brief, if we except Africa and the temperate parts of Australia, there is no large tract of country in the world that does not produce Cypripediums; and few authorities doubt that a larger acquaintance with those realms will bring them under the rule. We have a species in England, *C. calceolus,* by no means insignificant; it can be purchased from the dealers, but it is almost extinct in this country now. America furnishes a variety of species; which ought to be hardy. They will bear a frost below zero, but our winter damp is intolerable. Mr. Godseff tells me that he has seen *C. spectabile* growing like any water-weed in the bogs of New Jersey, where it is frozen hard, roots and all, for several months of the year; but very few survive the season in this country, even if protected. Those fine specimens so common at our spring shows are imported in the dry state. From the United States also we get the charming *C. candidum, C. parviflorum, C. pubescens,* and many more less important. Canada and Siberia furnish *C. guttatum, C. macranthum,* and others. I saw in Russia, and brought home, a magnificent species, tall and stately, bearing a great golden flower, which is not known "in the trade;" but they all rotted gradually. Therefore I do not recommend these fine outdoor varieties, [Pg 83]which the inexperienced are apt to think so easy. At the same cost others may be bought, which, coming from the highlands of hot countries, are used to a moderate damp in winter.

Foremost of these, perhaps the oldest of cool orchids in cultivation, is *C. insigne*, from Nepal. Everyone knows its original type, which has grown so common that I remarked a healthy pot at a window-garden exhibition some years ago in Westminster. One may say that this, the early and familiar form, has no value at present, so many fine varieties have been introduced. A reader may form a notion of the difference when I state that a small plant of exceptional merit sold for thirty guineas a short time ago—it was *C. insigne*, but glorified. This ranks among the fascinations of orchid culture. You may buy a lot of some common kind, imported, at a price representing coppers for each individual, and among them may appear, when they come to bloom, an eccentricity which sells for a hundred pounds or more. The experienced collector has a volume of such legends. There is another side to the question, truly, but it does not personally interest the class which I address. To make a choice among numberless stories of this sort, we may take the instance of *C. Spicerianum*.

[Pg 84]It turned up among a quantity of *Cypripedium insigne* in the greenhouse of Mrs. Spicer, a lady residing at Twickenham. Astonished at the appearance of this swan among her ducks, she asked Mr. Veitch to look at it. He was delighted to pay seventy guineas down for such a prize. Cypripediums propagate easily, no more examples came into the market, and for some years this lovely species was a treasure for dukes and millionaires. It was no secret that the precious novelty came from Mrs. Spicer's greenhouse; but to call on a strange lady and demand how she became possessed of a certain plant is not a course of action that commends itself to respectable business men. The circumstances gave no clue. Messrs. Spicer were and are large manufacturers of paper; there is no visible connection betwixt paper and Indian orchids. By discreet inquiries, however, it was ascertained that one of the lady's sons had a tea-plantation in Assam. No more was needed. By the next mail Mr. Forstermann started for that vague destination, and in process of time reached Mr. Spicer's bungalow. There he asked for "a job." None could be found for him; but tea-planters are hospitable, and the stranger was invited to stop a day or two. But he could not lead the conversation towards orchids—perhaps [Pg 85]because his efforts were too clever, perhaps because his host took no interest in the subject. One day, however, Mr. Spicer's manager invited him to go shooting, and casually remarked "we shall pass the spot where I found those orchids they're making such a fuss about at home." Be sure Mr. Forstermann was alert that morning! Thus put upon the track, he discovered quantities of it, bade the tea-planter adieu, and went to work; but in the very moment of triumph a tiger barred the way, his coolies bolted, and nothing would persuade them to go further. Mr. Forstermann was no shikari, but he felt himself called upon to uphold the cause of science and the honour of England at this juncture. In great agitation he went for that feline, and, in short, its skin still adorns Mrs. Sander's drawing-room. Thus it happened that on a certain Thursday a small pot of *C. Spicerianum* was sold, as usual, for sixty guineas at Stevens's; on the Thursday following all the world could buy fine plants at a guinea.

Cypripedium is the favourite orchid of the day. It has every advantage, except, to my perverse mind—brilliancy of colour. None show a whole tone; even the lovely *C. niveum* is not pure white. My views, however, find no backing. At all other [Pg 86]points the genus deserves to be a favourite. In the first place, it is the most interesting of all orchids to science. [3] Then its endless variations of form, its astonishing oddities, its wide range of hues, its easy culture, its readiness to hybridize and to ripen seed, the certainty, by comparison, of rearing the proceeds, each of these merits appeals to one or other of orchid-growers. Many of the species which come from torrid lands, indeed, are troublesome, but with such we are not concerned. The cool varieties will do well anywhere, provided they receive water enough in summer, and not too little in winter. I do not speak of the American and Siberian classes, which are nearly hopeless for the amateur, nor of the Hong-Kong *Cypripedium purpuratum*, a very puzzling example.

On the roll of martyrs to orchidology, Mr. Pearce stands high. To him we owe, among many fine things, the hybrid Begonias which are becoming such favourites for bedding and other purposes. He discovered the three original types, parents of the innumerable "garden flowers" now on sale—*Begonia Pearcii*, *B. Veitchii*, and *B. Boliviensis*. It was his great luck, and great honour, to find *Masdeval-*

lia Veitchii—so long, so often, so labori [Pg 87]ously searched for from that day to this, but never even heard of. To collect another shipment of that glorious orchid, Mr. Pearce sailed for Peru, in the service, I think, of Mr. Bull. Unhappily—for us all as well as for himself—he was detained at Panama. Somewhere in those parts there is a magnificent Cypripedium with which we are acquainted only by the dried inflorescence, named *planifolium*. The poor fellow could not resist this temptation. They told him at Panama that no white man had returned from the spot, but he went on. The Indians brought him back, some days or weeks later, without the prize; and he died on arrival.

Oncidiums also are a product of the New World exclusively; in fact, of the four classes most useful to amateurs, three belong wholly to America, and the fourth in great part. I resist the temptation to include Masdevallia, because that genus is not so perfectly easy as the rest; but if it be added, nine-tenths, assuredly, of the plants in our cool house come from the West. Among the special merits of the Oncidium is its colour. I have heard thoughtless persons complain that they are "all yellow;" which, as a statement of fact, is near enough to the truth, for about three-fourths may be so described roughly. But this dispensation is [Pg 88]another proof of Nature's kindly regard for the interests of our science. A clear, strong, golden yellow is the colour that would have been wanting in our cool houses had not the Oncidium supplied it. Shades of lemon and buff are frequent among Odontoglossums, but, in a rough, general way of speaking, they have a white ground. Masdevallias give us scarlet and orange and purple; Lycastes, green and dull yellow; Sophronitis, crimson; Mesospinidium, rose, and so forth. Blue must not be looked for. Even counting the new Utricularia for an orchid, as most people do, there are, I think, but five species that will live among us at present, in all the prodigious family, showing this colour; and every one of them is very "hot." Thus it appears that the Oncidium fills a gap—and how gloriously! There is no such pure gold in the scheme of the universe as it displays under fifty shapes wondrously varied. Thus—*Oncidium macranthum!* one is continually tempted to exclaim, as one or other glory of the orchid world recurs to mind, that it is the supreme triumph of floral beauty. I have sinned thus, and I know it. Therefore, let the reader seek an opportunity to behold *O. macranthum,* and judge for himself. But it seems to me that Nature gives us a hint. As though proudly conscious what a [Pg 89]marvel it will unfold, this superb flower often demands nine months to perfect itself. Dr. Wallace told me of an instance in his collection where eighteen months elapsed from the appearance of the spike until the opening of the first bloom. But it lasts a time proportionate.

Oncidium macranthum.
Reduced to One Sixth

Nature forestalled the dreams of æsthetic colourists when she designed *Oncidium macranthum*. Thus, and not otherwise, would the thoughtful of them arrange a "harmony" in gold and bronze; but Nature, with characteristic indifference to the fancies of mankind, hid her *chef-d'œuvre* in the wilds of Ecuador. Hardly less striking, however, though perhaps less beautiful, are its sisters of the "small-lipped" species — *Onc. serratum*, *O. superbiens*, and *O. sculptum*. This last is rarely seen. As with others of its class, the spike grows very long, twelve feet perhaps, if it were allowed to stretch. The flowers are small comparatively, clear bronze-brown, highly polished, so closely and daintily frilled round the edges that a fairy goffering-iron could not give more regular effects, and outlined by a narrow band of gold. *Onc. serratum* has a much larger bloom, but less compact, rather fly-away indeed, its sepals widening gracefully from a narrow neck. Excessively curious is the disposition of the petals, which close their tips to form [Pg 90]a circle of brown and gold around the column. The purpose of this extraordinary arrangement — unique among orchids, I believe — will be discovered one day, for purpose there is, no doubt; to judge by analogy, it may be supposed that the insect upon which *Onc. serratum* depends for fertilization likes to stand upon this ring while thrusting its proboscis into the nectary. The fourth of these fine species, *Onc. superbiens*, ranks among the grandest of flowers — knowing its own value, it rarely consents to "oblige;" the dusky green sepals are margined with yellow, petals white, clouded with pale purple, lip very small, of course, purple, surmounted by a great golden crest.

Most strange and curious is *Onc. fuscatum*, of which the shape defies description. Seen from the back, it shows a floriated cross of equal limbs; but in front the nethermost is hidden by a spreading lip, very large proportionately. The prevailing tint is a dun-purple, but each arm has a broad white tip. Dun-purple, also, is the centre of the labellum, edged with a distinct band of lighter hue, which again, towards the margin, becomes white. These changes of tone are not gradual, but as clear as a brush could make them. Botanists must long to dissect this extraordinary flower, but the opportunity seldom occurs. It is desperately puzzling [Pg 91]to understand how nature has packed away the component parts of its inflorescence, so as to resolve them into four narrow arms and a labellum. But the colouring of this plant is not always dull. In the small Botanic Garden at Florence, by Santa Maria Maggiore, I remarked with astonishment an *Onc. fuscatum*, of which the lip was scarlet-crimson and the other tints bright to match. That collection is admirably grown, but orchids are still scarce in Italy. The Society did not know what a prize it had secured by chance.

The genus Oncidium has, perhaps, more examples of a startling combination in hues than any other—but one must speak thoughtfully and cautiously upon such points.

I have not to deal with culture, but one hint may be given. Gardeners who have a miscellaneous collection to look after, often set themselves against an experiment in orchid-growing because these plants suffer terribly from green-fly and other pests, and will not bear "smoking." To keep them clean and healthy by washing demands labour for which they have no time. This is a very reasonable objection. But though the smoke of tobacco is actual ruination, no plant whatever suffers from the steam thereof. An ingenious Frenchman has invented and patented in England [Pg 92]lately a machine called the Thanatophore, which I confidently recommend. It can be obtained from Messrs. B.S. Williams, of Upper Holloway. The Thanatophore destroys every insect within reach of its vapour, excepting, curiously enough, scaly-bug, which, however, does not persecute cool orchids much. The machine may be obtained in different sizes through any good ironmonger.

To sum up: these plants ask nothing in return for the measureless enjoyment they give but light, shade from the summer sun, protection from the winter frost, moisture—and brains.

I am allowed to print a letter which bears upon several points to which I have alluded. It is not cheerful reading for the enthusiast. He will be apt to cry, "Would that the difficulties and perils were infinitely graver—so grave that the collecting grounds might have a rest for twenty years!"

January 19th, 1893.

Dear Sir,

I have received your two letters asking for *Cattleya Lawrenceana, Pancratium Guianense,* and *Catasetum pileatum.* Kindly excuse my answering your letters only to-day. But I have been [Pg 93]away in the interior, and on my return was sick, besides other business taking up my time; I was unable to write until to-day. Now let me give you some information concerning orchid-collecting in this colony. Six or seven years ago, just when the gold industry was starting, very few people ever ventured in the far interior. Boats, river-hands, and Indians could be hired at ridiculously low prices, and travelling and bartering paid; wages for Indians being about a shilling per day, and all found; the same for river-hands. Captains and boat-swains to pilot the boat through the rapids up and down for sixty-four cents a day. To-day you have got to pay sixty-four to eighty cents per day for Indians and river-hands. Captains and boatswains, $2 the former, and $1:50 the latter per day, and then you often cannot get them. Boat-hire used to be $8 to $10 for a big boat for three to four months; to-day $5, $6, and $7 per day, and all through the rapid development of the gold industry. As you can calculate twenty-five days' river travel to get within reach of the Savannah lands, you can reckon what the expenses must be, and then again about five to seven days coming down the river, and a couple of days to lay over. Then you must count two trips like this, one to bring you up, and one to bring you [Pg 94]down three months after, when you return with your collection. Besides this, you run the risk of losing your boat in the rapids either way, which happens not very unfrequently either going or coming; and we have not only to record the loss of several boats with goods, etc., every month, but generally to record the loss of life; only two cases happening last month, in one case seven, in the other twelve men losing their lives. Besides, river-hands and blacks will not go further than the boats can travel, and nothing will induce them to go among the Indians, being afraid of getting poisoned by Inds. (Kaiserimas) or strangled. So you have to rely utterly on Indians, which you often cannot get, as the district of Roraima is very poorly inhabited, and most of the Indians died by smallpox and measles breaking out among them four years ago, and those that survived left the district, and you will find whole districts nearly uninhabited. About five years ago I went up with Mr. Osmers to Roraima, but he broke down before we reached the Savannah. He lay there for a week, and I gave him up; he recovered, however, and dragged himself into the Savannah near Roraima, about three days distant from it, where I left him. Here we found and made a splendid collection of about 3000 first-class plants of different kinds.

[Pg 95]While I was going up to Roraima, he stayed in the Savannah, still too sick to go further. At Roraima I collected everything except *Catt. Lawrenceana*, which was utterly rooted out already by former collectors. On my return to Osmers' camp, I found him more dead than alive, thrown down by a new attack of sickness; but not alone that, I also found him abandoned by most of our Indians, who had fled on account of the Kanaima having killed three of their number. So Mr. Osmers—who got soon better—and I, made up our baskets with plants, and made everything ready. Our Indians returning partly, I sent him ahead with as many loads as we could carry, I staying behind with the rest of baskets of plants. Had all our Indians come back, we would have been all right, but this not being the case I had to stay until the Indians returned and fetched me off. After this we got back all right. This was before the sickness broke out among the Indians.

Last year I went up with Mr. Kromer, who met me going up-river while I was coming down. So I joined him. We got up all right to the river's head, but here our troubles began, as we got only about eight Indians to go on with us who had worked in the gold-diggings, and no others could be had, the district being abandoned. We had to [Pg 96]pay them half a dollar a day to carry loads. So we pushed on, carrying part of our loads, leaving the rest of our cargo behind, until we reached the Savannah, when we had to send them back several times to get the balance of our goods. From the time we reached the Savannah we were starving, more or less, as we could procure only very little provisions. We hunted all about for *Catt. Lawrenceana*, and got only about 1500 or so, it growing only here and there. At Roraima we did not hunt at all, as the district is utterly rubbed out by the Indians. We were about fourteen days at Roraima and got plenty of *Utricularia Campbelliana*, *U. Humboldtii*, and *U. montana*. Also *Zygopetalum*, *Cyp. Lindleyanum*, *Oncidium nigratum* (only fifty—very rare now), *Cypripedium Schomburgkianum*, *Zygopetalum Burkeii*, and in fact, all that is to be found on and about Roraima, except the *Cattleya Lawrenceana*. Also plenty others, as Sobralia, Liliastrum, etc. So our collection was not a very great one; we had the hardest trouble now through the want of Indians to carry the loads. Besides this, the rainy weather set in and our loads suffered badly for all the care we took of them. Besides, the Indians got disagreeable, having to go back several times to bring the remaining baskets. Nevertheless, we got down as [Pg 97]far as the Curubing mountains. Up to this time we were more or less always starving. Arrived at the Curubing mountains, procured a scant supply of provisions, but lost nearly all of them in a small creek, and what was saved was spoiling under our eyes, it being then that the rainy season had fully started, drenching us from morning to night. It took us nine days to get our loads over the mountain, where our boat was to reach us to take us down river. And we were for two and a half days entirely without food. Besides the plants being damaged by stress of weather, the Indians had opened the baskets and thrown partly the loads away, not being able to carry the heavy soaked-through baskets over the mountains, so making us lose the best of our plants.

Arrived at our landing we had to wait for our boat, which arrived a week later in consequence of the river being high, and, of course, short of provisions. Still, we got away with what we had of our loads until we reached the first gold places kept by a friend of mine, who supplied us with food. Thereafter we started for town. Halfway, at Kapuri falls (one of the most dangerous), we swamped down over a rock, and so we lost some of our things; still saved all our plants, though they lay for a few hours under water with the boat. [Pg 98] After this we reached town in safety. So after coming home we found, on packing up, that we had only about 900 plants, that is, *Cattleya Lawrenceana*, of which about one-third good, one-third medium, and one-third poor quality. This trip took us about three and a half months, and cost over 2500 dollars. Besides, I having poisoned my leg on a rotten stump which I run up in my foot, lay for four months suffering terrible pain.

You will, of course, see from this that orchid-hunting is no pleasure, as you of course know, but what I want to point out to you is that *Cattleya Lawrenceana* is very rare in the interior now.

The river expenses fearfully high, in fact, unreasonably high, on account of the gold-digging. Labourers getting 64 c. to $1.00 per day, and all found. No Indians to be got, and those that you can get at ridiculous prices, and getting them, too, by working on places where they build and thatch houses and clear the ground from underbrush, and as huntsmen for gold-diggers. Even if Mr. Kromer had succeeded to get 3000 or 4000 fine *Cattleya Lawrenceana*, it would have been of no value to us, as we could not have got anybody to carry them to the river where a boat could reach. Besides this, I also must tell you that there is a license to be paid out here if you want to collect orchids, [Pg 99]amounting to $100, which Mr. Kromer had to pay, and also an export tax duty of 2 cents per piece. So that orchid collecting is made a very expensive affair. Besides its success being very doubtful, even if a man is very well acquainted with Indian life and has visited the Savannah reaches year after year. We spent something over $2500 to $2900, including Mr. Kromer's and Steigfer's passage out, on our last expedition.

If you want to get any *Lawrenceana*, you will have to send yourself, and as I said before, the results will be very doubtful. As far as I myself am concerned, I am interested besides my baking business, in the gold-diggings, and shall go up to the Savannah in a few months. I can give you first-class references if you should be willing to send an expedition, and we could come to some arrangement; at least, you would save the expenses of the passage of one of your collectors. I may say that I am quite conversant with the way of packing orchids and handling them as well for travel as shipment.

Kindly excuse, therefore, my lengthy letter and its bad writing. And if you should be inclined to go in for an expedition, just send me a list of what you require, and I will tell you whether the plants are found along the route of travel and in the [Pg 100] Savannah visited; as, for instance, *Catt. superba* does not grow at all in the district where *Catt. Lawrenceana* is to be found, but far further south.

Before closing, I beg you to let me know the prices of about twenty-five of the best of and prettiest South American orchids, which I want for my own collection, as *Catt. Medellii, Catt. Trianæ, Odontoglossum crispum, Miltonia vexillaria, Catt. labiata*, &c.

I shall await your answer as soon as possible, and send you a list by last mail of what is to be got in this colony.

We also found on our last visit something new—a very large bulbed Oncidium, or may be Catasetum, on the top of Roraima, where we spent a night, but got only two specimens, one of which got lost, and the other one I left in the hands of Mr. Rodway, but so we tried our best. It decayed, having been too seriously damaged to revive and flower, and so enable us to see what it was, it not being in flower when found.

Awaiting your kind reply,
Yours truly,

Seyler.

P.S.—If you should send out one of your collectors, or require any information, I shall be glad to give it.

[Pg 101]One of the most experienced collectors, M. Oversluys, writes from the Rio de Yanayacca, January, 1893:—

"Here it is absolutely necessary that one goes himself into the woods ahead of the peons, who are quite cowards to enter the woods; and not altogether without reason, for the larger part of them get sick here, and it is very hard to enter—nearly impenetrable and full of insects, which make fresh-coming people to get cracked and mad. I have from the wrist down not a place to put in a shilling piece which is not a wound, through the very small red spider and other insects. Also my people are the same. Of the five men I took out, two have got fever already, and one ran back. To-morrow I expect other peons, but not a single one from Mengobamba. It is a trouble to get men who will come into the woods, and I cannot have more than eight or ten to work with, because when I should not be continually behind them or ahead they do nothing. It is not a question of money to do good here, but merely luck and the way one treats people. The peons come out less for their salaries than for good and plenty of food, which is very difficult to find in these scarce times....

"The plants are here one by one, and we have got but one tree with three plants. They are on [Pg 102]the highest and biggest trees, and these must be cut down with axes. Below are all shrubs, full of climbers and lianas about a finger thick. Every step must be cut to advance, and the ground cleared below the high trees in order to spy the branches. It is a very difficult job. Nature has well protected this Cattleya.... Nobody can like this kind of work."

The poor man ends abruptly, "I will write when I can—the mosquitos don't leave me a moment."

FOOTNOTES:

[2] See a letter at p. 92.

[3] *Vide* "Orchids and Hybridizing," *infra*, p. 210.

[Pg 103]

WARM ORCHIDS.

By the expression "warm" we understand that condition which is technically known as "intermediate." It is waste of time to ask, at this day, why a Latin combination should be employed when there is an English monosyllable exactly equivalent; we, at least, will use our mother-tongue. Warm orchids are those which like a minimum temperature, while growing, of 60°; while resting, of 55°. As for the maximum, it signifies little in the former case, but in the latter—during the months of rest—it cannot be allowed to go beyond 60°, for any length of time, without mischief. These conditions mean, in effect, that the house must be warmed during nine months of the twelve in this realm of England. "Hot" orchids demand a fire the whole year round—saving a few very rare nights when the Briton swelters in tropical discomfort. Upon this dry subject of temperature, however, I would add one word of encouragement for those who are not willing to pay [Pg 104] a heavy bill for coke. The cool-house, in general, requires a fire, at night, until June 1. Under that condition, if it face the south, in a warm locality, very many genera and species classed as intermediate should be so thoroughly started before artificial heat is withdrawn that they will do excellently, unless the season be unusual.

Warm orchids come from a sub-tropic region, or from the mountains of a hotter climate, where their kinsfolk dwelling in the plains defy the thermometer; just as in sub-tropic lands warm species occupy the lowlands, while the heights furnish Odontoglossums and such lovers of a chilly atmosphere. There are, however, some warm Odontoglossums, notable among them *O. vexillarium*, which botanists class with the Miltonias. This species is very fashionable, and I give it the place of honour; but not, in my own view, for its personal merits. The name is so singularly appropriate that one would like to hear the inventor's reasons for transfiguring it. *Vexillum* we know, and *vexillarius*, but *vexillarium* goes beyond my Latin. However, it is an intelligible word, and those acquainted with the appearance of "regimental colours" in Old Rome perceive its fitness at a glance. The flat bloom seems to hang suspended from its centre, just as the *vexillum* figures in bas-relief—on the Arch of Antoninus, for [Pg 105]example. To my mind the colouring is insipid, as a rule, and the general effect stark—fashion in orchids, as in other things, has little reference to taste. I repeat with emphasis, *as a rule*, for some priceless specimens are no less than astounding in their blaze of colour, the quintessence of a million uninteresting blooms. The poorest of these plants have merit, no doubt, for those who can accommodate giants. They grow fast and big. There are specimens in this country a yard across, which display a hundred and fifty or two hundred flowers open at the same time for months. A superb show they make, rising over the pale sea-green foliage, four spikes perhaps from a single bulb. But this is a beauty of general effect, which must not be analyzed, as I think.

Odontoglossum vexillarium is brought from Colombia. There are two forms: the one—small, evenly red, flowering in autumn—was discovered by Frank Klaboch, nephew to the famous Roezl, on the Dagua River, in Antioquia. For eight years he persisted in despatching small quantities to Europe, though every plant died; at length a safer method of transmission was found, but simultaneously poor Klaboch himself succumbed. It is an awful country—perhaps the wettest under the sun. Though a favourite hunting-ground of [Pg 106]collectors now—for Cattleyas of value come from hence, besides this precious Odontoglot—there are still no means of transport, saving Indians and canoes. *O. vexillarium* would not be thought costly if buyers knew how rare it is, how expensive to get, and how terribly difficult to bring home. Forty thousand pieces were despatched to Mr. Sander in one consignment—he hugged himself with delight when three thousand proved to have some trace of vitality.

Mr. Watson, Assistant Curator at Kew, recalls an amusing instance of the value and the mystery attached to this species so late as 1867. In that year Professor Reichenbach described it for the first time. He tells how a friend lent him the bloom upon a negative promise under five heads—"First, not to show it to any one else; (2) not to speak much about it; (3) not to take a drawing of it; (4) not to have a photograph made; (5) not to look oftener than three times at it." By-the-bye, Mr. Watson gives the credit of the first discovery to the late Mr. Bowman; but I venture to believe that my account is exact—in reference to the Antioquia variety, at least.

The other form occurs in the famous district of Frontino, about two hundred and fifty miles due north of the first habitat, and shows—*savants* [Pg 107] would add "of course"—a striking difference. In the geographical distinctions of species will be found the key to whole volumes of mystery that perplex us now. I once saw three Odontoglossums ranged side by side, which even an expert would pronounce mere varieties of the same plant if he were not familiar with them—*Od. Williamsi*, *Od. grande*, and *Od. Schlieperianum*. The middle one everybody knows, by sight at least, a big, stark, spread-eagle flower, gamboge yellow mottled with red-brown, vastly effective in the mass, but individually vulgar. On one side was *Od. Williamsi*, essentially the same in flower and bulb and growth, but smaller; opposite stood *Od. Schlieperianum*, only to be distinguished as smaller still. But both these latter rank as species. They are separated from the common type, *O. grande*, by nearly ten degrees of latitude and ten degrees of longitude, nor—we might almost make an affidavit—do any intermediate forms exist in the space between; and those degrees are sub-tropical, by so much more significant than an equal distance in our zone. Instances of the same class and more surprising are found in many genera of orchid.

The Frontino *vexillarium* grows "cooler," has a much larger bloom, varies in hue from purest white to deepest red, and flowers in May or June. [Pg 108] The most glorious of these things, however, is *O. vex. superbum*, a plant of the greatest rarity, conspicuous for its blotch of deep purple in the centre of the lip, and its little dot of the same on each wing. Doubtless this is a natural hybrid betwixt the Antioquia form and *Odontoglossum Roezlii*, which is its neighbour. The chance of finding a bit of *superbum* in a bundle of the ordinary kind lends peculiar excitement to a sale of these plants. Such luck first occurred to Mr. Bath, in Stevens' Auction Rooms. He paid half-a-crown for a very weakly fragment, brought it round, flowered it, and received a prize for good gardening in the shape of seventy-two pounds, cheerfully paid by Sir Trevor Lawrence for a plant unique at that time. I am reminded of another little story. Among a great number of *Cypripedium insigne* received at St. Albans, and "established," Mr. Sander noted one presently of which the flower-stalk was yellow instead of brown, as is usual. Sharp eyes are a valuable item of the orchid-grower's stock-in-trade, for the smallest peculiarity among such "sportive" objects should not be neglected. Carefully he put the yellow stalk aside—the only one among thousands, one might say myriads, since *C. insigne* is one of our oldest and commonest orchids, [Pg 109]and it never showed this phenomenon before. In due course the flower opened, and proved to be all golden! Mr. Sander cut his plant in two, sold half for seventy-five pounds to a favoured customer, and the other half, publicly, for one hundred guineas. One of the purchasers has divided his plant now and sold two bits at 100 guineas. Another piece was bought back by Mr. Sander, who wanted it for hybridizing, at 250 guineas—not a bad profit for the buyer, who has still two plants left. Another instance occurs to me while I write—such legends of shrewdness worthily rewarded fascinate a poor journalist who has the audacity to grow orchids. Mr. Harvey, solicitor, of Liverpool, strolling through the houses at St. Albans on July 24, 1883, remarked a plant of *Lælia anceps*, which had the ring-mark on its pseudo-bulb much higher up than is usual. There might be some meaning in that eccentricity, he thought, paid two guineas for the little thing, and on December 1, 1888, sold it back to Mr. Sander for 200*l*. It proved to be *L. a. Amesiana*, the grandest form of *L. anceps* yet discovered—rosy white, with petals deeply splashed; thus named after F.L. Ames, an American amateur. Such pleasing opportunities might arise for you or me any day.

[Pg 110]The first name that arises to most people in thinking of warm orchids is Cattleya, and naturally. The genus Odontoglossum alone has more representatives under cultivation. Sixty species of Cattleya are grown by amateurs who pay special attention to these plants; as for the number of "varieties" in a single species, one boasts forty, another thirty, several pass the round dozen. They are exclusively American, but they flourish over all the enormous space between Mexico and the Argentine Republic. The genus is not a favourite of my own, for somewhat of the same reason which qualifies my regard for *O. vexillarium*. Cattleyas are so obtrusively beautiful, they have such great flowers, which they thrust upon the eye with such assurance of admiration! Theirs is a style of effect—I refer to the majority—which may be called infantine; such as an intelligent and tasteful child might conceive if he had no fine sense of colour, and were too young to distinguish a showy from a charming form. But I say no more.

The history of Orchids long established is uncertain, but I believe that the very first Cattleya which appeared in Europe was *C. violacea Loddigesi*, imported by the great firm whose name it bears, to which we owe such a heavy debt. Two [Pg 111]years later came *C. labiata*, of which more must be said; then *C. Mossiæ*, from Caraccas; fourth, *C. Trianæ* named after Colonel Trian, of Tolima, in the United States of Colombia. Trian well deserved immortality, for he was a native of that secluded land—and a botanist! It is a natural supposition that his orchid must be the commonest of weeds in its home; seeing how all Europe is stocked with it, and America also, rash people might say there are millions in cultivation. But it seems likely that *C. Trianæ* was never very frequent, and at the present time assuredly it is so scarce that collectors are not sent after it. Probably the colonel, like many other *savants*, was an excellent man of business, and he established "a corner" when he saw the chance. *C. Mossiæ* stands in the same situation—or indeed worse; it can scarcely be found now. These instances convey a serious warning. In seventy years we have destroyed the native stock of two orchids, both so very free in propagating that they have an exceptional advantage in the struggle for existence. How long can rare species survive, when the demand strengthens and widens year by year, while the means of communication and transport become easier over all the world? Other instances will be mentioned in their place.

[Pg 112]Island species are doomed, unless, like *Lœlia elegans*, they have inaccessible crags on which to find refuge. It is only a question of time; but we may hope that Governments will interfere before it is too late. Already Mr. Burbidge has suggested that "some one" who takes an interest in orchids should establish a farm, a plantation, here and there about the world, where such plants grow naturally, and devote himself to careful hybridization on the spot. "One might make as much," he writes, "by breeding orchids as by breeding cattle, and of the two, in the long run, I should prefer the orchid farm." This scheme will be carried out one day, not so much for the purpose of hybridization as for plain "market-gardening;" and the sooner the better.

The prospect is still more dark for those who believe—as many do—that no epiphytal orchid under any circumstances can be induced to establish itself permanently in our greenhouses as it does at home. Doubtless, they say, it is possible to grow them and to flower them, by assiduous care, upon a scale which is seldom approached under the rough treatment of Nature. But they are dying from year to year, in spite of appearances. That it is so in a few cases can hardly be denied; but, seeing how many plants which have not changed [Pg 113]hands since their establishment, twenty or thirty or forty years ago, have grown continually bigger and finer, it seems much more probable that our ignorance is to blame for the loss of those species which suddenly collapse. Sir Trevor Lawrence observed the other day: "With regard to the longevity of orchids, I have one which I know to have been in this country for more than fifty years, probably even twenty years longer than that— *Renanthera coccinea*." The finest specimens of Cattleya in Mr. Stevenson Clarke's houses have been "grown on" from small pieces imported twenty years ago. If there were more collections which could boast, say, half a century of uninterrupted attention, we should have material for forming a judgment; as a rule, the dates of purchase or establishment were not carefully preserved till late years.

But there is one species of Cattleya which must needs have seventy years of existence in Europe, since it had never been rediscovered till 1890. When we see a pot of *C. labiata*, the true, autumn-flowering variety, more than two years old, we know that the very plant itself must have been established about 1818, or at least its immediate parent—for no seedling has been raised to public knowledge. [4]

[Pg 114]In avowing a certain indifference to Cattleyas, I referred to the bulk, of course. The most gorgeous, the stateliest, the most imperial of all flowers on this earth, is *C. Dowiana*—unless it be *C. aurea*, a "geographical variety" of the same. They dwell a thousand miles apart at least, the one in Colombia, the other in Costa Rica; and neither occurs, so far as is known, in the great intervening region. Not even a connecting link has been discovered; but the Atlantic coast of Central America is hardly explored, much less examined. In my time it was held, from Cape Camarin to Chagres, by independent tribes of savages—not independent in fact alone, but in name also. The Mosquito Indians are recognized by Europe as free; the Guatusos kept a space of many hundred miles from which no white man had returned; when I was in those parts, the Talamancas, though not so unfriendly, were only known by the report of adventurous pedlars. I made an attempt—comparatively spirited—to organize an exploring party for the benefit of the Guatusos, but no single volunteer answered our advertisements in San José de Costa Rica; I have lived to congratulate myself on that disappointment. Since my day a road has been cut through their wilds to Limon, certain luckless Britons having found the money for a railway; [Pg 115]but an engineer who visited the coast but two years ago informs me that no one ever wandered into "the bush." Collectors have not been there, assuredly. So there may be connecting links between *C. Dowiana* and *C. aurea* in that vast wilderness, but it is quite possible there are none.

Words could not picture the glory of these marvels. In each the scheme of colour is yellow and crimson, but there are important modifications. Yellow is the ground all through in *Cattleya aurea*—sepals, petals, and lip; unbroken in the two former, in the latter superbly streaked with crimson. But *Cattleya Dowiana* shows crimson pencillings on its sepals, while the ground colour of the lip is crimson, broadly lined and reticulated with gold. Imagine four of these noble flowers on one stalk, each half a foot across! But it lies beyond the power of imagination.

C. Dowiana was discovered by Warscewicz about 1850, and he sent home accounts too enthusiastic for belief. Steady-going Britons utterly refused to credit such a marvel—his few plants died, and there was an end of it for the time. I may mention an instance of more recent date, where the eye-witness of a collector was flatly rejected at home. Monsieur St. Leger, residing at Asuncion, the capital of Paraguay, wrote a warm description of [Pg 116]an orchid in those parts to scientific friends. The account reached England, and was treated with derision. Monsieur St. Leger, nettled, sent some dried flowers for a testimony; but the mind of the Orchidaceous public was made up. In 1883 he brought a quantity of plants and put them up at auction; nobody in particular would buy. So those reckless or simple or trusting persons who invested a few shillings in a bundle had all the fun to themselves a few months afterwards, when the beautiful *Oncidium Jonesianum* appeared, to confound the unbelieving. It must be added, however, that orchid-growers may well become an incredulous generation. When their judgment leads them wrong we hear of it, the tale is published, and outsiders mock. But these gentlemen receive startling reports continually, honest enough for the most part. Much experience and some loss have made them rather cynical when a new wonder is announced. The particular case of Monsieur St. Leger was complicated by the extreme resemblance which the foliage of *Onc. Jonesianum* bears to that of *Onc. cibolletum,* a species almost worthless. Unfortunately the beautiful thing declines to live with us—as yet.

Cattleya Dowiana was rediscovered by Mr. Arce, when collecting birds: it must have been a [Pg 117]grand moment for Warscewicz when the horticultural world was convulsed by its appearance in bloom. *Cattleya aurea* had no adventures of this sort. Mr. Wallis found it in 1868 in the province of Antioquia, and again on the west bank of the Magdalena; but it is very rare. This species is persecuted in its native home by a beetle, which accompanies it to Europe not infrequently—in the form of eggs, no doubt. A more troublesome alien is the fly which haunts *Cattleya Mendellii,* and for a long time prejudiced growers against that fine species, until, in fact, they had made a practical and rather costly study of its habits. An experienced grower detects the presence of this enemy at a glance. It pierces an "eye" — a back one in general, happily — and deposits an egg in the very centre. Presently this growth begins to swell in a manner that delights the ingenuous horticulturist, until he remarks that its length does not keep pace with its breadth. But one remedy has yet been discovered — cutting off any suspected growth. We understand now that *C. Mendellii* is as safe to import as any other species, unless it be gathered at the wrong time. [5]

[Pg 118]Among the most glorious, rarest, and most valuable of Cattleyas is *C. Hardyana*, doubtless a natural hybrid of *C. aurea* with *C. gigas Sanderiana*. Few of us have seen it — two-hundred-guinea plants are not common spectacles. It has an immense flower, rose-purple; the lip purple-magenta, veined with gold. *Cattleya Sanderiana* offers an interesting story. Mr. Mau, one of Mr. Sander's collectors, was despatched to Bogota in search of *Odontoglossum crispum*. While tramping through the woods, he came across a very large Cattleya at rest, and gathered such pieces as fell in his way — attaching so little importance to them, however, that he did not name the matter in his reports. Four cases Mr. Mau brought home with his stock of Odontoglossums, which were opened in due course of business. We can quite believe that it was one of the stirring moments of Mr. Sander's life. The plants bore many dry specimens of last year's inflorescence, displaying such extraordinary size as proved the variety to be new; and there is no large Cattleya of indifferent colouring. To receive a plant of that character unannounced, undescribed, is an experience without parallel for half a century. Mr. Mau was sent back by next mail to secure every fragment he could find. Meantime, those in hand [Pg 119]were established, and Mr. Brymer, M.P., bought one — Mr. Brymer is immortalized by the Dendrobe which bears his name. The new Cattleya proved kindly, and just before Mr. Mau returned with some thousands of its like Mr. Brymer's purchase broke into bloom. That must have been another glorious moment for Mr. Sander, when the great bud unfolded, displaying sepals and petals of the rosiest, freshest, softest pink, eleven inches across; and a crimson labellum exquisitely shown up by a broad patch of white on either side of the throat. Mr. Brymer was good enough to lend his specimen for the purpose of advertisement, and Messrs. Stevens enthusiastically fixed a green baize partition across their rooms as a background for the wondrous novelty. What excitement reigned there on the great day is not to be described. I have heard that over 2000*l*. was taken in the room.

Most of the Cattleyas with which the public is familiar—*Mossiæ*, *Trianæ*, *Mendellii*, and so forth—have white varieties; but an example absolutely pure is so uncommon that it fetches a long price. Loveliest of these is *C. Skinneri alba*. For generations, if not for ages, the people of Costa Rica have been gathering every morsel they can find, and planting it upon the roofs of their mud-built [Pg 120]churches. Roezl and the early collectors had a "good time," buying these semi-sacred flowers from the priests, bribing the parishioners to steal them, or, when occasion served, playing the thief themselves. But the game is nearly up. Seldom now can a piece of *Cat. Skinneri alba* be obtained by honest means, and when a collector arrives guards are set upon the churches that still keep their decoration. No plant has ever been found in the forest, we understand.

It is just the same case with *Lælia anceps alba*. The genus Lælia is distinguished from Cattleya by a peculiarity to be remarked only in dissection; its pollen masses are eight as against four. To my taste, however, the species are more charming on the whole. There is *L. purpurata*. Casual observers always find it hard to grasp the fact that orchids are weeds in their native homes, just like foxgloves and dandelions with us. In this instance, as I have noted, they flatly refuse to believe, and certainly "upon the face of it" their incredulity is reasonable.

Lælia purpurata falls under the head of hot orchids. *L. anceps*, however, is not so exacting; many people grow it in the cool house when they can expose it there to the full blaze of sunshine. In its commonest form it is divinely beautiful. [Pg 121] I have seen a plant in Mr. Eastey's collection with twenty-three spikes, the flowers all open at once. Such a spectacle is not to be described in prose. But when the enthusiast has rashly said that earth contains no more ethereal loveliness, let him behold *L. a. alba*, the white variety. The dullest man I ever knew, who had a commonplace for all occasions, found no word in presence of that marvel. Even the half-castes of Mexico who have no soul, apparently, for things above horseflesh and cockfights, and love-making, reverence this saintly bloom. The Indians adore it. Like their brethren to the south, who have tenderly removed every plant of *Cattleya Skinneri alba* for generations unknown, to set upon their churches, they collect this supreme effort of Nature and replant it round their huts. So thoroughly has the work been done in either case that no single specimen was ever seen in the forest. Every one has been bought from the Indians, and the supply is exhausted; that is to say, a good many more are known to exist, but very rarely now can the owner be persuaded to part with one. The first example reached England nearly half a century ago, sent probably by a native trader to his correspondent in this country; but, as was usual at that time, the circumstances are doubtful. It found its way, somehow, to Mr. [Pg 122] Dawson, of Meadowbank, a famous collector, and by him it was divided. Search was made for the treasure in its home, but vainly; travellers did not look in the Indian gardens. No more arrived for many years. Mr. Sander once conceived a fine idea. He sent one of his collectors to gather *Lœlia a. alba* at the season when it is in bud, with an intention of startling the universe by displaying a mass of them in full bloom; they were still more uncommon then than now, when a dozen flowering plants is still a show of which kings may be proud. Mr. Bartholomeus punctually fulfilled his instructions, collected some forty plants with their spikes well developed; attached them to strips of wood which he nailed across shallow boxes, and shipped them to San Francisco. Thence they travelled by fast train to New York, and proceeded without a moment's delay to Liverpool on board the *Umbria*; it was one of her first trips. All went well. Confidently did Mr. Sander anticipate the sensation when a score of those glorious plants were set out in full bloom upon the tables. But on opening the boxes he found every spike withered. The experiment is so tempting that it has been essayed once more, with a like result. The buds of *Lœlia anceps* will not stand sea air.

Catasetums do not rank as a genus among our [Pg 123]beauties; in fact, saving *C. pileatum*, commonly called *C. Bungerothi*, and *C. barbatum*, I think of none, at this moment, which are worthy of attraction on that ground. *C. fimbriatum*, indeed, would be lovely if it could be persuaded to show itself. I have seen one plant which condescended to open its spotted blooms, but only one. No orchids, however, give more material for study; on this account Catasetum was a favourite with Mr. Darwin. It is approved also by unlearned persons who find relief from the monotony of admiration as they stroll round in observing its acrobatic performances. The "column" bears two horns; if these be touched, the pollen-masses fly as if discharged from a catapult. *C. pileatum*, however, is very handsome, four inches across, ivory white, with a round well in the centre of its broad lip, which makes a theme for endless speculation. The daring eccentricities of colour in this class of plant have no stronger example than *C. callosum*, a novelty from Caraccas, with inky brown sepals and petals, brightest orange column, labellum of verdigris-green tipped with orange to match.

Schomburgkias are not often seen. Having a boundless choice of fine things which grow and flower without reluctance, the practical gardener [Pg 124]gets irritated in these days when he finds a plant beyond his skill. It is a pity, for the Schomburgkias are glorious things—in especial *Sch. tibicinis*. No description has done it justice, and few are privileged to speak as eye-witnesses. The clustering flowers hang down, sepals and petals of dusky mauve, most gracefully frilled and twisted, encircling a great hollow labellum which ends in a golden drop. That part of the cavity which is visible between the handsome incurved wings has bold stripes of dark crimson. The species is interesting, too. It comes from Honduras, where the children use its great hollow pseudo-bulbs as trumpets— whence the name. At their base is a hole—a touch-hole, as we may say, the utility of which defies our botanists. Had Mr. Belt travelled in those parts, he might have discovered the secret, as in the similar case of the Bullthorn, one of the *Gummiferæ*. The great thorns of that bush have just such a hole, and Mr. Belt proved by lengthy observations that it is designed, to speak roughly, for the ingress of an ant peculiar to that acacia, whose duty it is to defend the young shoots—*vide* Belt's "Naturalist in Nicaragua," page 218. Importers are too well aware that *Schomburgkia tibicinis* also is inhabited by an ant of singular ferocity, for it survives the voyage, [Pg 125]and rushes forth to battle when the case is opened. We may suppose that it performs a like service.

Dendrobiums are "warm" mostly; of the hot species, which are many, and the cool, which are few, I have not to speak here. But a remark made at the beginning of this chapter especially applies to Dendrobes. If they be started early, so that the young growths are well advanced by June 1; if the situation be warm, and a part of the house sunny—if they be placed in that part without any shade till July, and freely syringed—with a little extra attention many of them will do well enough. That is to say, they will make such a show of blossom as is mighty satisfactory in the winter time. We must not look for "specimens," but there should be bloom enough to repay handsomely the very little trouble they give. Among those that may be treated so are *D. Wardianum, Falconeri, crassinode, Pierardii, crystallinum, Devonianum*—sometimes—and *nobile*, of course. Probably there are more, but these I have tried myself.

Dendrobium Wardianum, at the present day, comes almost exclusively from Burmah—the neighbourhood of the Ruby Mines is its favourite habitat. But it was first brought to England from [Pg 126] Assam in 1858, when botanists regarded it as a form of *D. Falconeri*. This error was not so strange as its seems, for the Assamese variety has pseudo-bulbs much less sturdy than those we are used to see, and they are quite pendulous. It was rather a lively business collecting orchids in Burmah before the annexation. The Roman Catholic missionaries established there made it a source of income, and they did not greet an intruding stranger with warmth—not genial warmth, at least. He was forbidden to quit the town of Bhamo, an edict which compelled him to employ native collectors—in fact, coolies—himself waiting helplessly within the walls; but his reverend rivals, having greater freedom and an acquaintance with the language, organized a corps of skirmishers to prowl round and intercept the natives returning with their loads. Doubtless somebody received the value when they made a haul, but who, is uncertain perhaps—and the stranger was disappointed, anyhow. It may be believed that unedifying scenes arose—especially on two or three occasions when an agent had almost reached one of the four gates before he was intercepted. For the hapless collector—having nothing in the world to do—haunted those portals all day long, flying from one to the other in hope to see "somebody [Pg 127]coming." Very droll, but Burmah is a warm country for jests of the kind. Thus it happened occasionally that he beheld his own discomfiture, and rows ensued at the Mission-house. At length Mr. Sander addressed a formal petition to the Austrian Archbishop, to whom the missionaries owed allegiance. He received a sympathetic answer, and some assistance.

From the Ruby Mines also comes a Dendrobium so excessively rare that I name it only to call the attention of employés in the new company. This is *D. rhodopterygium*. Sir Trevor Lawrence has or had a plant, I believe; there are two or three at St. Albans; but the lists of other dealers will be searched in vain. Sir Trevor Lawrence had also a scarlet species from Burmah; but it died even before the christening, and no second has yet been found. Sumatra furnishes a scarlet Dendrobe, *D. Forstermanni*, but it again is of the utmost rarity. Baron Schroeder boasts three specimens—which have not yet flowered, however. From Burmah comes *D. Brymerianum*, of which the story is brief, but very thrilling if we ponder it a moment. For the missionaries sent this plant to Europe without a description—they had not seen the bloom, doubtless—and it sold cheap enough. We may fancy Mr. Brymer's emotion, therefore, when the striking [Pg 128]flower opened. Its form is unique, though some other varieties display a long fringe—as that extraordinary object, *Nanodes Medusæ*, and also *Brassavola Digbyana*, which is exquisitely lovely sometimes. In the case of *D. Brymerianum* the bright yellow lip is split all round, for two-thirds of its expanse, into twisted filaments. We may well ask what on earth is Nature's purpose in this eccentricity; but it is a question that arises every hour to the most thoughtless being who grows orchids.

Dendrobium Brymerianum.
Reduced To One Fourth

Everybody knows *Dendrobium nobile* so well that it is not to be discussed in prose; something might be done in poetry, perhaps, by young gentlemen who sing of buttercups and daisies, but the rhyme would be difficult. *D. nobile nobilius*, however, is by no means so common—would it were! This glorified form turned up among an importation made by Messrs. Rollisson. They propagated it, and sold four small pieces, which are still in cultivation. But the troubles of that renowned firm, to which we owe so great a debt, had already begun. The mother-plant was neglected. It had fallen into such a desperate condition when Messrs. Rollisson's plants were sold, under a decree in bankruptcy, that the great dealers refused to bid for what should have been a little gold-mine. A casual market-gardener hazarded thirty shillings, brought it round so far that he could establish a number of young plants, and sold the parent for forty pounds at last. There are, however, several fine varieties of *D. nobile* more valuable than *nobilius*. *D. n. Sanderianum* resembles that form, but it is smaller and darker. Albinos have been found; Baron Schroeder has a beautiful example. One appeared at Stevens' Rooms, announced as the single instance in cultivation—which is not quite the fact, but near enough for the auction-room, perhaps. It also was imported originally by Mr. Sander, with *D. n. Sanderianum*. Biddings reached forty-three pounds, but the owner would not deal at the price. Albinos are rare among the Dendrobes.

D. nobile Cooksoni was the *fons et origo* of an unpleasant misunderstanding. It turned up in the collection of Mr. Lange, distinguished by a reversal of the ordinary scheme of colour. There is actually no end to the delightful vagaries of these plants. If people only knew what interest and pleasing excitement attends the inflorescence of an imported orchid—one, that is, which has not bloomed before in Europe—they would crowd the auction-rooms in which every strange face is marked now. There are books enough to inform them, certainly; but who reads an Orchid Book? Even the enthusiast only consults it.

Dendrobium nobile Cooksoni, then, has white tips to petal and sepal; the crimson spot keeps its place; and the inside of the flower is deep red—an inversion of the usual colouring. Mr. Lange could scarcely fail to observe this peculiarity, but he seems to have thought little of it. Mr. Cookson, paying him a visit, was struck, however—as well he might be—and expressed a wish to have the plant. So the two distinguished amateurs made an exchange. Mr. Cookson sent a flower at once to Professor Reichenbach, who, delighted and enthusiastic, registered it upon the spot under the name of the gentleman from whom he received it. Mr. Lange protested warmly, demanding that his discovery should be called, after his residence, *Heathfield-sayeanum*. But Professor Reichenbach drily refused to consider personal questions; and really, seeing how short is life, and how long *Dendrobium nobile Heathfield*, &c., true philanthropists will hold him justified.

We may expect wondrous Dendrobes from New Guinea. Some fine species have already arrived, and others have been sent in the dried inflorescence. Of *D. phalænopsis Schroederi* I have spoken elsewhere. There is *D. Goldiei*; a variety of *D. superbiens*—but much larger. There is *D. Albertesii*, snow-white; *D. Broomfieldianum*, curiously like *Lælia anceps alba* in its flower—which is to say that it must be the loveliest of all Dendrobes. But this species has a further charm, almost incredible. The lip in some varieties is washed with lavender blue, in some with crimson! Another is nearly related to *D. bigibbum*, but much larger, with sepals more acute. Its hue is a glorious rosy-purple, deepening on the lip, the side lobes of which curl over and meet, forming a cylindrical tube, while the middle lobe, prolonged, stands out at right angles, veined with very dark purple; this has just been named *D. Statterianum*. It has upon the disc an elevated, hairy crest, like *D. bigibbum*, but instead of being white as always, more or less, in that instance, the crest of the new species is dark purple. I have been particular in describing this noble flower, because very, very few have beheld it. Those who live will see marvels when the Dutch and German portions of New Guinea are explored.

Recently I have been privileged to see another, the most impressive to my taste, of all the lovely genus. It is called *D. atro-violaceum*. The stately flowers hang down their heads, reflexed like a "Turban

Lily," ten or a dozen on a spike. The colour is ivory-white, with a faintest tinge of green, [Pg 132]and green spots are dotted all over. The lobes of the lip curl in, making half the circumference of a funnel, the outside of which is dark violet-blue; with that fine colour the lip itself is boldly striped. They tell me that the public is not expected to "catch on" to this marvel. It hangs its head too low, and the contrast of hues is too startling. If that be so, we multiply schools of art and County Council lectures perambulate the realm, in vain. The artistic sense is denied us.

Madagascar also will furnish some astonishing novelties; it has already begun, in fact—with a vengeance. Imagine a scarlet Cymbidium! That such a wonder existed has been known for some years, and three collectors have gone in search of it; two died, and the third has been terribly ill since his return to Europe—but he won the treasure, which we shall behold in good time. Those parts of Madagascar which especially attract botanists must be death-traps indeed! M. Léon Humblot tells how he dined at Tamatave with his brother and six compatriots, exploring the country with various scientific aims. Within twelve months he was the only survivor. One of these unfortunates, travelling on behalf of Mr. Cutler, the celebrated naturalist of Bloomsbury Street, to find butterflies and birds, shot at a native idol, as the report [Pg 133]goes. The priests soaked him with paraffin, and burnt him on a table—perhaps their altar. M. Humblot himself has had awful experiences. He was attached to the geographical survey directed by the French Government, and ten years ago he found *Phajus Humblotii* and *Phajus tuberculosus* in the deadliest swamps of the interior. A few of the bulbs gathered lived through the passage home, and caused much excitement when offered for sale at Stevens' Auction Rooms. M. Humblot risked his life again, and secured a great quantity for Mr. Sander, but at a dreadful cost. He spent twelve months in the hospital at Mayotte, and on arrival at Marseilles with his plants the doctors gave him no hope of recovery. *P. Humblotii* is a marvel of beauty—rose-pink, with a great crimson labellum exquisitely frilled, and a bright green column.

Everybody who knows his "Darwin" is aware that Madagascar is the chosen home of the Angræcums. All, indeed, are natives of Africa, so far as I know, excepting the delightful *A. falcatum*, which comes, strangely enough, from Japan. One cannot but suspect, under the circumstances, that this species was brought from Africa ages ago, when the Japanese were enterprising seamen, and has been acclimatized by those skilful horti [Pg 134]culturists. It is certainly odd that the only "cool" Aerides—the only one found, I believe, outside of India and the Eastern Tropics—also belongs to Japan, and a cool Dendrobe, *A. arcuatum*, is found in the Transvaal; and I have reason to hope that another or more will turn up when South Africa is thoroughly searched. A pink Angræcum, very rarely seen, dwells somewhere on the West Coast; the only species, so far as I know, which is not white. It bears the name of M. Du Chaillu, who found it—he has forgotten where, unhappily. I took that famous traveller to St. Albans in the hope of quickening his recollection, and I fear I bored him afterwards with categorical inquiries. But all was vain. M. Du Chaillu can only recall that once on a time, when just starting for Europe, it occurred to him to run into the bush and strip the trees indiscriminately. Mr. Sander was prepared to send a man expressly for this Angræcum. The exquisite *A. Sanderianum* is a native of the Comorro Islands. No flower could be prettier than this, nor more deliciously scented—when scented it is! It grows in a climate which travellers describe as Paradise, and, in truth, it becomes such a scene. Those who behold young plants with graceful garlands of snowy bloom twelve to twenty inches long are [Pg 135]prone to fall into raptures; but imagine it as a long-established specimen appears just now at St Albans, with racemes drooping two and a half feet from each new growth, clothed on either side with flowers like a double train of white long-tailed butterflies hovering! *A. Scottianum* comes from Zanzibar, discovered, I believe, by Sir John Kirk; *A. caudatum*, from Sierra Leone. This latter species is the nearest rival of *A. sesquipedale*, showing "tails" ten inches long. Next in order for this characteristic detail rank *A. Leonis* and *Kotschyi*—the latter rarely grown—with seven-inch "tails;" *Scottianum* and *Ellisii* with six-inch; that is to say, they ought to show such dimensions respectively. Whether they fulfil their promise depends upon the grower.

With the exceptions named, this family belongs to Madagascar. It has a charming distinction, shared by no other genus which I recall, save, in less degree, Cattleya—every member is attractive. But I must concentrate myself on the most striking—that which fascinated Darwin. In the first place it should be pointed out that *savants* call this plant *Æranthus sesquipedalis*, not *Angræcum*—a fact useful to know, but unimportant to ordinary mortals. It was discovered by the Rev. Mr. Ellis, and sent home alive, nearly thirty years ago; [Pg 136]but civilized mankind has not yet done wondering at it. The stately growth, the magnificent green-white flowers, command admiration at a glance, but the "tail," or spur, offers a problem of which the thoughtful never tire. It is commonly ten inches long, sometimes fourteen inches, and at home, I have been told, even longer; about the thickness of a goose-quill, hollow, of course, the last inch and a half filled with nectar. Studying this appendage by the light of the principles he had laid down, Darwin ventured on a prophecy which roused special mirth among the unbelievers. Not only the abnormal length of the nectary had to be considered; there was, besides, the fact that all its honey lay at the base, a foot or more from the orifice. Accepting it as a postulate that every detail of the apparatus must be equally essential for the purpose it had to serve, he made a series of experiments which demonstrated that some insect of Madagascar—doubtless a moth—must be equipped with a proboscis long enough to reach the nectar, and at the same time thick enough at the base to withdraw the pollinia—thus fertilizing the bloom. For, if the nectar had lain so close to the orifice that moths with a proboscis of reasonable length and thickness could get at it, they would drain the cup without touching the pollinia. [Pg 137] Darwin never proved his special genius more admirably than in this case. He created an insect beyond belief, as one may say, by the force of logic; and such absolute confidence had he in his own syllogism that he declared, "If such great moths were to become extinct in Madagascar, assuredly this Angræcum would become extinct." I am not aware that Darwin's fine argument has yet been clinched by the discovery of that insect. But cavil has ceased. Long before his death a sphinx moth arrived from South Brazil which shows a proboscis between ten and eleven inches long—very nearly equal, therefore, to the task of probing the nectary of *Angræcum sesquipidale*. And we know enough of orchids at this time to be absolutely certain that the Madagascar species must exist.

FOOTNOTES:

[4] *Vide* "The Lost Orchid," *infra*, p. 173.

[5] I have learned by a doleful experience that this fly, commonly called "the weavil," is quite at home on *Lælia purpurata*; in fact, it will prey on any Cattleya.

[Pg 138]

HOT ORCHIDS.

In former chapters I have done my best to show that orchid culture is no mystery. The laws which govern it are strict and simple, easy to define in books, easily understood, and subject to few exceptions. It is not with Odontoglossums and Dendrobes as with roses — an intelligent man or woman needs no long apprenticeship to master their treatment. Stove orchids are not so readily dealt with; but then, persons who own a stove usually keep a gardener. Coming from the hot lowlands of either hemisphere, they show much greater variety than those of the temperate and sub-tropic zones; there are more genera, though not so many species, and more exceptions to every rule. These, therefore, are not to be recommended to all householders. Not everyone indeed is anxious to grow plants which need a minimum night heat of 60° in winter, 70° in summer, and cannot dispense with fire the whole year round.

The hottest of all orchids probably is *Peristeria elata*, [Pg 139] the famous "Spirito Santo," flower of the Holy Ghost. The dullest soul who observes that white dove rising with wings half spread, as in the very act of taking flight, can understand the frenzy of the Spaniards when they came upon it. Rumours of Peruvian magnificence had just reached them at Panama — on the same day, perhaps — when this miraculous sign from heaven encouraged them to advance. The empire of the Incas did not fall a prey to that particular band of ruffians, nevertheless. *Peristeria elata* is so well known that I would not dwell upon it, but an odd little tale rises to my mind. The great collector Roezl was travelling homeward, in 1868, by Panama. The railway fare to Colon was sixty dollars at that time, and he grudged the money. Setting his wits to work, Roezl discovered that the company issued tickets from station to station at a very low price for the convenience of its employés. Taking advantage of this system, he crossed the isthmus for five dollars—such an advantage it is in travelling to be an old campaigner! At one of the intermediate stations he had to wait for his train, and rushed into the jungle of course. *Peristeria* abounded in that steaming swamp, but the collector was on holiday. To his amazement, however, he found, side by side with it, a Masde [Pg 140]vallia—that genus most impatient of sunshine among all orchids, flourishing here in the hottest blaze! Snatching up half a dozen of the tender plants with a practised hand, he brought them safe to England. On the day they were put up to auction news of Livingstone's death arrived, and in a flash of inspiration Roezl christened his novelty *M. Livingstoniana*. Few, indeed, even among authorities, know where that rarest of Masdevallias has its home; none have reached Europe since. A pretty flower it is—white, rosy tipped, with yellow "tails." And it dwells by the station of Culebras, on the Panama railway.

Of genera, however, doubtless the Vandas are hottest; and among these, *V. Sanderiana* stands first. It was found in Mindanao, the most southerly of the Philippines, by Mr. Roebelin when he went thither in search of the red Phalœnopsis, as will be told presently. *Vanda Sanderiana* is a plant to be described as majestic rather than lovely, if we may distinguish among these glorious things. Its blooms are five inches across, pale lilac in their ground colour, suffused with brownish yellow, and covered with a network of crimson brown. Twelve or more of such striking flowers to a spike, and four or five spikes upon a plant make a wonder indeed. But, to view matters prosaically, *Vanda* [Pg 141] *Sanderiana* is "bad business." It is not common, and it grows on the very top of the highest trees, which must be felled to secure the treasure; and of those gathered but a small proportion survive. In the first place, the agent must employ natives, who are paid so much per plant, no matter what the size—a bad system, but they will allow no change. It is evidently their interest to divide any "specimen" that will bear cutting up; if the fragments bleed to death, they have got their money meantime. Then, the Manilla steamers call at Mindanao only once a month. Three months are needed to get together plants enough to yield a fair profit. At the end of that time a large proportion of those first gathered will certainly be doomed—Vandas have no pseudo-bulbs to sustain their strength. Steamers run from Manilla to Singapore every fortnight. If the collector be fortunate he may light upon a captain willing to receive his packages; in that case he builds structures of bamboo on deck, and spends the next fortnight in watering, shading, and ventilating his precious *trouvailles*, alternately. But captains willing to receive such freight must be waited for too often. At Singapore it is necessary to make a final overhauling of the plants—to their woeful diminution. This done, troubles recom [Pg 142]mence. Seldom will the captain of a mail steamer accept that miscellaneous cargo. Happily, the time of year is, or ought to be, that season when tea-ships arrive at Singapore. The collector may reasonably hope to secure a passage in one of these, which will carry him to England in thirty-five days or so. If this state of things be pondered, even without allowance for accident, it will not seem surprising that *V. Sanderiana* is a costly species. The largest piece yet secured was bought by Sir Trevor Lawrence at auction for ninety guineas. It had eight stems, the tallest four feet high. No consignment has yet returned a profit, however.

The favoured home of Vandas is Java. They are noble plants even when at rest, if perfect—that is, clothed in their glossy, dark green leaves from base to crown. If there be any age or any height at which the lower leaves fall of necessity, I have not been able to identify it. In Mr. Sander's collection, for instance, there is a giant plant of *Vanda suavis*, eleven growths, a small thicket, established in 1847. The tallest stem measures fifteen feet, and every one of its leaves remain. They fall off easily under bad treatment, but the mischief is reparable at a certain sacrifice. The stem may be cut through and the crown re [Pg 143]planted, with leaves perfect; but it will be so much shorter, of course. The finest specimen I ever heard of is the *V. Lowii* at Ferrières, seat of Baron Alphonse de Rothschild, near Paris. It fills the upper part of a large greenhouse, and year by year its twelve stems produce an indefinite number of spikes, eight to ten feet long, covered with thousands of yellow and brown blooms. [6] Vandas inhabit all the Malayan Archipelago; some are found even in India. The superb *V. teres* comes from Sylhet; from Burmah also. This might be called the floral cognizance of the house of Rothschild. At Frankfort, Vienna, Ferrières, and Gunnersbury little meadows of it are grown—that is, the plants flourish at their own sweet will, uncumbered with pots, in houses devoted to them. Rising from a carpet of palms and maidenhair, each crowned with its drooping garland of rose and crimson and cinnamon-brown, they make a glorious show indeed. A pretty little coincidence was remarked when the Queen paid a visit to Waddesdon the other day. *V. teres* first bloomed in Europe at Syon House, and a small spray was sent to the young Princess, unmarried then and uncrowned. The incident recurred to memory when Baron Ferdinand de Rothschild chose this [Pg 144]same flower for the bouquet presented to Her Majesty; he adorned the luncheon table therewith besides. This story bears a moral. The plant of which one spray was a royal gift less than sixty years ago has become so far common that it may be used in masses to decorate a room. Thousands of unconsidered subjects of Her Majesty enjoy the pleasure which one great duke monopolized before her reign began. There is matter for an essay here. I hasten back to my theme.

V. teres is not such a common object that description would be superfluous. It belongs to the small class of climbing orchids, delighting to sun itself upon the rafters of the hottest stove. If this habit be duly regarded, it is not difficult to flower by any means, though gardeners who do not keep pace with their age still pronounce it a hopeless rebel. Sir Hugh Low tells me that he clothed all the trees round Government House at Pahang with *Vanda teres*, planting its near relative, *V. Hookeri*, more exquisite still, if that were possible, in a swampy hollow. His servants might gather a basket of these flowers daily in the season. So the memory of the first President for Pahang will be kept green. A plant rarely seen is *V. limbata* from the island of Timor—dusky [Pg 145]yellow, the tip purple, outlined with white, formed like a shovel.

I may cite a personal reminiscence here, in the hope that some reader may be able to supply what is wanting. In years so far back that they seem to belong to a "previous existence," I travelled in Borneo, and paid a visit to the antimony-mines of Bidi. The manager, Mr. Bentley, showed me a grand tapong-tree at his door from which he had lately gathered a "blue orchid,"—we were desperately vague about names in the jungle at that day, or in England for that matter. In a note published on my return, I said, "As Mr. Bentley described it, the blossoms hung in an azure garland from the bough, more gracefully than art could design." This specimen is, I believe, the only one at present known, and both Malays and Dyaks are quite ignorant of such a flower! What was this? There is no question of the facts. Mr. Bentley sent the plant, a large mass to the chairman of the Company, and it reached home in fair condition. I saw the warm letter, enclosing cheque for 100*l.*, in which Mr. Templar acknowledged receipt. But further record I have not been able to discover. One inclines to assume that a blue orchid which puts forth a "garland" of bloom must be a Vanda. The description might [Pg 146]be applied to *V. cœrulea*, but that species is a native of the Khasya hills; more appropriately, as I recall Mr. Bentley's words, to *V. cœrulescens*, which, however, is Burmese. Furthermore, neither of these would be looked for on the branch of a great tree. Possibly someone who reads this may know what became of Mr. Templar's specimen.

Both the species of Renanthera need great heat. Among "facts not generally known" to orchid-growers, but decidedly interesting for them, is the commercial habitat, as one may say, of *R. coccinea*. The books state correctly that it is a native of Cochin China. Orchids coming from such a distance must needs be withered on arrival. Accordingly, the most experienced horticulturist who is not up to a little secret feels assured that all is well when he beholds at the auction-room or at one of the small dealer's a plant full of sap, with glossy leaves and unshrivelled roots. It must have been in cultivation for a year at the very least, and he buys with confidence. Too often, however, a disastrous change sets in from the very moment his purchase reaches home. Instead of growing it falls back and back, until in a very few weeks it has all the appearance of a newly-imported piece. The explanation is [Pg 147]curious. At some time, not distant, a quantity of *R. coccinea* must have found its way to the neighbourhood of Rio. There it flourishes as a weed, with a vigour quite unparalleled in its native soil. Unscrupulous persons take advantage of this extraordinary accident. From a country so near and so readily accessible they can get plants home, pot them up, and sell them, before the withering process sets in. May this revelation confound such knavish tricks! The moral is old—buy your orchids from one of the great dealers, if you do not care to "establish" them yourself.

R. coccinea is another of the climbing species, and it demands, even more urgently than *V. teres*, to reach the top of the house, where sunshine is fiercest, before blooming. Under the best conditions, indeed, it is slow to produce its noble wreaths of flower — deep red, crimson, and orange. Upon the other hand, the plant itself is ornamental, and it grows very fast. The Duke of Devonshire has some at Chatsworth which never fail to make a gorgeous show in their season; but they stand twenty feet high, twisted round birch-trees, and they have occupied their present quarters for half a century or near it. There is but one more species in the genus, so far as the unlearned know, but this, generally [Pg 148]recognized as *Vanda Lowii*, as has been already mentioned, ranks among the grand curiosities of botanic science. Like some of the Catasetums and Cycnoches, it bears two distinct types of flower on each spike, but the instance of *R. Lowii* is even more perplexing. In those other cases the differing forms represent male and female sex, but the microscope has not yet discovered any sort of reason for the like eccentricity of this Renanthera. Its proper inflorescence, as one may put it, is greenish yellow, blotched with brown, three inches in diameter, clothing a spike sometimes twelve feet long. The first two flowers to open, however — those at the base — present a strong contrast in all respects — smaller, of different shape, tawny yellow in colour, dotted with crimson. It would be a pleasing task for ingenious youth with a bent towards science to seek the utility of this arrangement.

Orchids are spreading fast over the world in these days, and we may expect to hear of other instances where a species has taken root in alien climes like *R. coccinea* in Brazil. I cannot cite a parallel at present. But Mr. Sander informs me that there is a growing demand for these plants in realms which have their own native orchids. We have an example in the letter which has been [Pg 149]already quoted. [7] Among customers who write to him direct are magnates of China and Siam, an Indian and a Javanese rajah. Orders are received — not unimportant, nor infrequent — from merchants at Calcutta, Singapore, Hong Kong, Rio de Janeiro, and smaller places, of course. It is vastly droll to hear that some of these gentlemen import species at a great expense which an intelligent coolie could gather for them in any quantity within a few furlongs of their go-down! But for the most part they demand foreigners.

The plants thus distributed will be grown in the open air; naturally they will seed; at least, we may hope so. Even *Angræcum sesquipedale*, of which I wrote in the preceding chapter, would find a moth able to impregnate it in South Brazil. Such species as recognize the conditions necessary for their existence will establish themselves. It is fairly safe to credit that in some future time, not distant, Cattleyas may flourish in the jungles of India, Dendrobiums on the Amazons, Phalœnopsis in the coast lands of Central America. Those who wish well to their kind would like to hasten that day.

Mr. Burbidge suggested at the Orchid Conference that gentlemen who have plantations in a [Pg 150]country suitable should establish a "farm," or rather a market-garden, and grow the precious things for exportation. It is an excellent idea, and when tea, coffee, sugar-cane, all the regular crops of the East and West Indies, are so depreciated by competition, one would think that some planters might adopt it. Perhaps some have; it is too early yet for results. Upon inquiry I hear of a case, but it is not encouraging. One of Mr. Sander's collectors, marrying when on service in the United States of Colombia, resolved to follow Mr. Burbidge's advice. He set up his "farm" and began "hybridizing" freely. No man living is better qualified as a collector, for the hero of this little tale is Mr. Kerbach, a name familiar among those who take interest in such matters; but I am not aware that he had any experience in growing orchids. To start with hybridizing seems very ambitious—too much of a short cut to fortune. However, in less than eighteen months Mr. Kerbach found it did not answer, for reasons unexplained, and he begged to be reinstated in Mr. Sander's service. It is clear, indeed, that the orchid-farmer of the future, in whose success I firmly believe, will be wise to begin modestly, cultivating the species he finds in his neighbourhood. It is not in our greenhouses alone that [Pg 151]these plants sometimes show likes and dislikes beyond explanation. For example, many gentlemen in Costa Rica—a wealthy land, and comparatively civilized—have tried to cultivate the glorious *Cattleya Dowiana*. For business purposes also the attempt has been made. But never with success. In those tropical lands a variation of climate or circumstances, small perhaps, but such as plants that subsist mostly upon air can recognize, will be found in a very narrow circuit. We say that Trichopilias have their home at Bogota. As a matter of fact, however, they will not live in the immediate vicinity of that town, though the woods, fifteen miles away, are stocked with them. The orchid-farmer will have to begin cautiously, propagating what he finds at hand, and he must not be hasty in sending his crop to market. It is a general rule of experience that plants brought from the forest and "established" before shipment do less well than those shipped direct in good condition, though the public, naturally, is slow to admit a conclusion opposed by *à priori* reasoning. The cause may be that they exhaust their strength in that first effort, and suffer more severely on the voyage.

I hear of one gentleman, however, who appears to be cultivating orchids with success. This is [Pg 152] Mr. Rand, dwelling on the Rio Negro, in Brazil, where he has established a plantation of *Hevia Brazilienses*, a new caoutchouc of the highest quality, indigenous to those parts. Some years ago Mr. Rand wrote to Mr. Godseff, at St. Albans, begging plants of *Vanda Sanderiana* and other Oriental species, which were duly forwarded. In return he despatched some pieces of a new Epidendrum, named in his honour *E. Randii*, a noble flower, with brown sepals and petals, the lip crimson, betwixt two large white wings. This and others native to the Rio Negro Mr. Rand is propagating on a large scale in shreds of bamboo, especially a white *Cattleya superba* which he himself discovered. It is pleasing to add that by latest reports all the Oriental species were thriving to perfection on the other side of the Atlantic.

Vandas, indeed, should flourish where *Cattleya superba* is at home, or anything else that loves the atmosphere of a kitchen on washing-day at midsummer. Though all the Cattleyas, or very nearly all, will "do" in an intermediate house, several prefer the stove. Of two among them, *C. Dowiana* and *C. aurea*, I spoke in the preceding chapter with an enthusiasm that does not bear repetition. *Cattleya guttata Leopoldi* grows upon [Pg 153]rocks in the little island of Sta. Catarina, Brazil, in company with *Lælia elegans* and *L. purpurata*. There the four dwelt in such numbers only twenty years ago that the supply was thought inexhaustible. It has come to an end already, and collectors no longer visit the spot. Cliffs and ravines which men still young can recollect ablaze with colour, are as bare now as a stone-quarry. Nature had done much to protect her treasures; they flourished mostly in places which the human foot cannot reach—*Lælia elegans* and *Cattleya g. Leopoldi* inextricably entwined, clinging to the face of lofty rocks. The blooms of the former are white and mauve, of the latter chocolate-brown, spotted with dark red, the lip purple. A wondrous sight that must have been in the time of flowering. It is lost now, probably for ever. Natives went down, suspended on a rope, and swept the whole circuit of the island, year by year. A few specimens remain in nooks absolutely inaccessible, but those happy mortals who possess a bit of *L. elegans* should treasure it, for more are very seldom forthcoming. *Lælia elegans Statteriana* is the finest variety perhaps; the crimson velvet tip of its labellum is as clearly and sharply-defined upon the snow-white surface as pencil could draw; it looks like [Pg 154]painting by the steadiest of hands in angelic colour. *C. g. Leopoldi* has been found elsewhere. It is deliciously scented. I observed a plant at St. Albans lately with three spikes, each bearing over twenty flowers; many strong perfumes there were in the house, but that overpowered them all. The *Lælia purpurata* of Sta. Catarina, to which the finest varieties in cultivation belong, has shared the same fate. It occupied boulders jutting out above the swamps in the full glare of tropic sunshine. Many gardeners give it too much shade. This species grows also on the mainland, but of inferior quality in all respects; curiously enough it dwells upon trees there, even though rocks be at hand, while the island variety, I believe, was never found on timber.

Another hot Cattleya of the highest class is *C. Acklandiæ* It belongs to the dwarf section of the genus, and inexperienced persons are vastly surprised to see such a little plant bearing two flowers on a spike, each larger than itself. They are four inches in diameter, petals and sepals chocolate-brown, barred with yellow, lip large, of colour varying from rose to purple. *C. Acklandiæ* is found at Bahia, where it grows side by side with *C. amethystoglossa*, also a charming species, very tall, leafless to the tip of its pseudo-bulbs. Thus [Pg 155]the dwarf beneath is seen in all its beauty. As they cling together in great masses the pair must make a flower-bed to themselves—above, the clustered spikes of *C. amethystoglossa*, dusky-lilac, purple-spotted, with a lip of amethyst; upon the ground the rich chocolate and rose of *C. Acklandiæ*.

Cattleya superba, as has been said, dwells also on the Rio Negro in Brazil; it has a wide range, for specimens have been sent from the Rio Meta in Colombia. This species is not loved by gardeners, who find it difficult to cultivate and almost impossible to flower, probably because they cannot give it sunshine enough. I have heard that Baron Hruby, a Hungarian enthusiast in our science, has no sort of trouble; wonders, indeed, are reported of that admirable collection, where all the hot orchids thrive like weeds. The Briton may find comfort in assuming that cool species are happier beneath his cloudy skies; if he be prudent, he will not seek to verify the assumption. The Assistant Curator of Kew assures us, in his excellent little work, "Orchids," that the late Mr. Spyers grew *C. superba* well, and he details his method. I myself have never seen the bloom. Mr. Watson describes it as five inches across, "bright rosy-purple suffused with white, very fragrant, lip with acute side lobes folding over the [Pg 156]column,"—making a funnel, in short—"the front lobe spreading, kidney-shaped, crimson-purple, with a blotch of white and yellow in front."

In the same districts with *Cattleya superba* grows *Galleandra Devoniana* under circumstances rather unusual. It clings to the very tip of a slender palm, in swamps which the Indians themselves regard with dread as the chosen home of fever and mosquitoes. It was discovered by Sir Robert Schomburgk, who compared the flower to a foxglove, referring especially, perhaps, to the graceful bend of its long pseudo-bulbs, which is almost lost under cultivation. The tube-like flowers are purple, contrasting exquisitely with a snow-white lip, striped with lilac in the throat.

Phalœnopsis, of course, are hot. This is one of our oldest genera which still rank in the first class. It was drawn and described so early as 1750, and a plant reached Messrs. Rollisson in 1838; they sold it to the Duke of Devonshire for a hundred guineas. Many persons regard Phalœnopsis as the loveliest of all, and there is no question of their supreme beauty, though not everyone may rank them first. They come mostly from the Philippines, but Java, Borneo, Cochin China, Burmah, even Assam contribute some species. Colonel Berkeley found *Ph. tetraspis*, [Pg 157]snow-white, and *Ph. speciosa*, purple, in the Andamans, when he was Governor of that settlement, clinging to low bushes along the mangrove creeks. So far as I know, all the species dwell within breath of the sea, as it may be put, where the atmosphere is laden with salt; this gives a hint to the thoughtful. Mr. Partington, of Cheshunt, who was the most renowned cultivator of the genus in his time, used to lay down salt upon the paths and beneath the stages of his Phalœnopsis house. Lady Howard de Walden stands first, perhaps, at the present day, and her gardener follows the same system. These plants, indeed, are affected, for good or ill, by influences too subtle for our perception as yet. Experiment alone will decide whether a certain house, or a certain neighbourhood even, is agreeable to their taste. It is a waste of money in general to make alterations; if they do not like the place they won't live there, and that's flat! It is probable that Maidstone, where Lady Howard de Walden resides, may be specially suited to their needs, but her ladyship's gardener knows how to turn a lucky chance to the best account. Some of his plants have ten leaves! — the uninitiated may think that fact grotesquely undeserving of a note of exclamation, but to explain would be too [Pg 158]technical. It may be observed that the famous Swan orchid, *Cycnoches chlorochilon*, flourishes at Maidstone as nowhere else perhaps in England.

Phalœnopsis were first introduced by Messrs. Rollisson, of Tooting, a firm that vanished years ago, but will live in the annals of horticulture as the earliest of the great importers. In 1836 they got home a living specimen of *Ph. amabilis*, which had been described, and even figured, eighty years before. A few months later the Duke of Devonshire secured *Ph. Schilleriana*. The late Mr. B.S. Williams told me a very curious incident relating to this species. It comes from the Philippines, and exacts a very hot, close atmosphere of course. Once upon a time, however, a little piece was left in the cool house at Holloway, and remained there some months unnoticed by the authorities. When at length the oversight was remarked, to their amaze this stranger from the tropics, abandoned in the temperate zone, proved to be thriving more vigorously than any of his fellows who enjoyed their proper climate! — so he was left in peace and cherished as a "phenomenon." Four seasons had passed when I beheld the marvel, and it was a picture of health and strength, flowering freely; but the reader is not advised to introduce a few Phalœ[Pg 159]nopsis to his Odontoglossums — not by any means. Mr. Williams himself never repeated the experiment. It was one of those delightfully perplexing vagaries which the orchid-grower notes from time to time.

There are rare species of this genus which will not be found in the dealers' catalogues, and amateurs who like a novelty may be pleased to hear some names. *Ph. Manni*, christened in honour of Mr. Mann, Director of the Indian Forest Department, is yellow and red; *Ph. cornucervi*, yellow and brown; *Ph. Portei*, a natural hybrid, of *Ph. rosea* and *Ph. Aphrodite*, white, the lip amethyst. It is found very, very rarely in the woods near Manilla. Above all, *Ph. Sanderiana*, to which hangs a little tale.

So soon as the natives of the Philippines began to understand that their white and lilac weeds were cherished in Europe, they talked of a scarlet variety, which thrilled listening collectors with joy; but the precious thing never came to hand, and, on closer inquiry, no responsible witness could be found who had seen it. Years passed by and the scarlet Phalœnopsis became a jest among orchidaceans. The natives persisted, however, and Mr. Sander found the belief so general, if shadowy, that when a service of coasting steamers [Pg 160]was established, he sent Mr. Roebelin to make a thorough investigation. His enterprise and sagacity were rewarded, as usual. After floating round for twenty-five years amidst derision, the rumour proved true in part. *Ph. Sanderiana* is not scarlet but purplish rose, a very handsome and distinct species.

To the same collector we owe the noblest of Aerides, *A. Lawrenciæ*, waxy white tipped with purple, and deep purple lip. Besides the lovely colouring it is the largest by far of that genus. Mr. Roebelin sent two plants from the Far East; he had not seen the flower, nor received any description from the natives. Mr. Sander grew them in equal ignorance for three years, and sent one to auction in blossom; it fell to Sir Trevor Lawrence's bid for 235 guineas.

Coelogene Pandurata.
Reduced to One Sixth

Many of the Cœlogenes classed as cool, which, indeed, rub along with Odontoglossums, do better in the stove while growing. *Cœl. cristata* itself comes from Nepaul, where the summer sun is terrible, and it covers the rocks most exposed. But I will only name a few of those recognized as hot. Amongst the most striking of flowers, exquisitely pretty also, is *Cœl. pandurata*, from Borneo. Its spike has been described by a person of fine fancy as resembling a row of glossy pea- [Pg 161]green frogs with black tongues, each three inches in diameter. The whole bloom is brilliantly green, but several ridges clothed with hairs as black and soft as velvet run down the lip, seeming to issue from a mouth. It is strange to see that a plant so curious, so beautiful, and so sweet should be so rarely cultivated; I own, however, that it is very unwilling to make itself at home with us. *Cœl. Dayana*, also a native of Borneo, one of our newest discoveries, is named after Mr. Day, of Tottenham. I may interpolate a remark here for the encouragement of poor but enthusiastic members of our fraternity. When Mr. Day sold his collection lately, an American "Syndicate" paid 12,000*l*. down, and the remaining plants fetched 12,000*l*. at auction; so, at least, the uncontradicted report goes. *Cœl. Dayana* is rare, of course, and dear, but Mr. Sander has lately imported a large quantity. The spike is three feet long sometimes, a pendant wreath of buff-yellow flowers broadly striped with chocolate. *Cœl. Massangeana*, from Assam, resembles this, but the lip is deep crimson-brown, with lines of yellow, and a white edge. Newest of all the Cœlogenes, and supremely beautiful, is *Cœl. Sanderiana*, imported by the gentleman whose name it bears. He has been called "The Orchid [Pg 162] King." This superb species has only flowered once in Europe as yet; Baron Ferdinand Rothschild is the happy man. Its snow-white blooms, six on a spike generally, each three inches across, have very dark brown stripes on the lip. It was discovered in Borneo by Mr. Forstermann, the same collector who happed upon the wondrous scarlet Dendrobe, mentioned in a former chapter. There I stated that Baron Schroeder had three pieces; this was a mistake unfortunately. Mr. Forstermann only secured three, of which two died on the journey. Baron Schroeder bought the third, but it has perished. No more can be found as yet.

Of Oncidiums there are many that demand stove treatment. The story of *Onc. splendidum* is curious. It first turned up in France some thirty years ago. A ship's captain sailing from St. Lazare brought half a dozen pieces, which he gave to his "owner," M. Herman. The latter handed them to MM. Thibaut and Ketteler, of Sceaux, who split them up and distributed them. Two of the original plants found their way to England, and they also appear to have been cut up. A legend of the King Street Auction Room recalls how perfervid competitors ran up a bit of *Onc. splendidum*, that had only one leaf, to thirty [Pg 163]guineas. The whole stock vanished presently, which is not surprising if it had all been divided in the same ruthless manner. From that day the species was lost until Mr. Sander turned his attention to it. There was no record of its habitat. The name of the vessel, or even of the captain, might have furnished a clue had it been recorded, for the shipping intelligence of the day would have shown what ports he was frequenting about that time. I could tell of mysterious orchids traced home upon indications less distinct. But there was absolutely nothing. Mr. Sander, however, had scrutinized the plant carefully, while specimens were still extant, and from the structure of the leaf he formed a strong conclusion that it must belong to the Central American flora; furthermore, that it must inhabit a very warm locality. In 1882 he directed one of his collectors, Mr. Oversluys, to look for the precious thing in Costa Rica. Year after year the search proceeded, until Mr. Oversluys declared with some warmth that *Onc. splendidum* might grow in heaven or in the other place, but it was not to be found in Costa Rica. But theorists are stubborn, and year after year he was sent back. At length, in 1882, riding through a district often explored, the collector found himself in a grassy plain, [Pg 164]dotted with pale yellow flowers. He had beheld the same many times, but his business was orchids. On this occasion, however, he chanced to approach one of the masses, and recognized the object of his quest. It was the familiar case of a man who overlooks the thing he has to find, because it is too near and too conspicuous. But Mr. Oversluys had excuse enough. Who could have expected to see an Oncidium buried in long grass, exposed to the full power of a tropic sun?

Oncidium Lanceanum is, perhaps, the hottest of its genus. Those happy mortals who can grow it declare they have no trouble, but unless perfectly strong and healthy it gets "the spot," and promptly goes to wreck. In the houses of the "New Plant and Bulb Company," at Colchester—now extinct—*Onc. Lanceanum* flourished with a vigour almost embarrassing, putting forth such enormous leaves, as it hung close to the glass, as made blinds quite superfluous at midsummer. But this was an extraordinary case. Certainly it is a glorious spectacle in flower—yellow, barred with brown; the lip violet. The spikes last a month in full beauty—sometimes two.

An Oncidium which always commands attention from the public and grateful regard from the devotee is *Onc. papilio*. Its strange form fascinated [Pg 165]the Duke of Devonshire, grandfather to the present, who was almost the first of our lordly amateurs, and tempted him to undertake the explorations which introduced so many fine plants to Europe.

The "Butterfly orchid" is so familiar that I do not pause to describe it. But imagine that most interesting flower all blue, instead of gold and brown! I have never been able to learn what was the foundation of the old belief in such a marvel. But the great Lindley went to his grave in unshaken confidence that a blue *papilio* exists. Once he thought he had a specimen; but it flowered, and his triumph had to be postponed. I myself heard of it two years back, and tried to cherish a belief that the news was true. A friend from Natal assured me that he had seen one on the table of the Director of the Gardens at Durban; but it proved to be one of those terrestrial orchids, so lovely and so tantalizing to us, with which South Africa abounds. Very slowly do we lengthen the catalogue of them in our houses. There are gardeners, such as Mr. Cook at Loughborough, who grow *Disa grandiflora* like a weed. Mr. Watson of Kew demonstrated that *Disa racemosa* will flourish under conditions easily secured. I had the good fortune [Pg 166]to do as much for *Disa Cooperi*, though not by my own skill. One supreme little triumph is mine, however. In very early days, when animated with the courage of utter ignorance, I bought eight bulbs of *Disa discolor*, and flowered them, every one! No mortal in Europe had done it before, nor has any tried since, I charitably hope, for a more rubbishing bloom does not exist. But there it was—*Ego feci*! And the specimen in the Herbarium at Kew bears my name.

But legends should not be disregarded when it is certain that they reach us from a native source. Some of the most striking finds had been announced long since by observant savages. I have told the story of *Phalœnopsis Sanderiana*. It was a Zulu who put the discoverer of the new yellow Calla on the track. The blue Utricularia had been heard of and discredited long before it was found—Utricularias are not orchids indeed, but only botanists regard the distinction. The natives of Assam persistently assert that a bright yellow Cymbidium grows there, of supremest beauty, and we expect it to turn up one day; the Malagasy describe a scarlet one. But I am digressing.

Epidendrums mostly will bear as much heat as can be given them while growing; all demand [Pg 167]more sunshine than they can get in our climate. Amateurs do not seem to be so well acquainted with the grand things of this genus as they should be. They distrust all imported Epidendrums. Many worthless species, indeed, bear a perplexing resemblance to the finest; so much so, that the most observant of authorities would not think of buying at the auction-room unless he had confidence enough in the seller's honesty to accept his description of a "lot." Gloriously beautiful, however, are some of those rarely met with; easy to cultivate also, in a sunny place, and not dear. *Epid. rhizophorum* has been lately rechristened *Epid. radicans*—a name which might be confined to the Mexican variety. For the plant recurs in Brazil, practically the same, but with a certain difference. The former grows on shrubs, a true epiphyte; the latter has its bottom roots in the soil, at foot of the tallest trees, and runs up to the very summit, perhaps a hundred and fifty feet. The flowers also show a distinction, but in effect they are brilliant orange-red, the lip yellow, edged with scarlet. Forty or fifty of them hanging in a cluster from the top of the raceme make a show to remember. Mr. Watson "saw a plant a few years ago, that bore eighty-six heads of flowers!" They last for three months. *Epid. prismatocarpum*, [Pg 168]also, is a lovely thing, with narrow dagger-like sepals and petals, creamy-yellow, spotted black, lip mauve or violet, edged with pale yellow.

Of the many hot Dendrobiums, Australia supplies a good proportion. There is *D. bigibbum*, of course, too well known for description; it dwells on the small islands in Torres Straits. This species flowered at Kew so early as 1824, but the plant died. Messrs. Loddiges, of Hackney, re-introduced it thirty years later. *D. Johannis*, from Queensland, brown and yellow, streaked with orange, the flowers curiously twisted. *D. superbiens*, from Torres Straits, rosy purple, edged with white, lip crimson. Handsomest of all by far is *D. phalœnopsis*. It throws out a long, slender spike from the tip of the pseudo-bulb, bearing six or more flowers, three inches across. The sepals are lance-shaped, and the petals, twice as broad, rosy-lilac, with veins of darker tint; the lip, arched over by its side lobes, crimson-lake in the throat, paler and striped at the mouth. It was first sent home by Mr. Forbes, of Kew Gardens, from Timor Laüt, in 1880. But Mr. Fitzgerald had made drawings of a species substantially the same, some years before, from a plant he discovered on the property of Captain Bloomfield, Balmain, in Queensland, nearly a thousand miles south of [Pg 169] Timor. Mr. Sander caused search to be made, and he has introduced Mr. Fitzgerald's variety under the name of *D. ph. Statterianum*. It is smaller than the type, and crimson instead of lilac.

Bulbophyllums rank among the marvels of nature. It is a point comparatively trivial that this genus includes the largest of orchids and, perhaps, the smallest.

B. Beccarii has leaves two feet long, eighteen inches broad. It encircles the biggest tree in one clasp of its rhizomes, which travellers mistake for the coil of a boa constrictor. Furthermore, this species emits the vilest stench known to scientific persons, which is a great saying. But these points are insignificant. The charm of Bulbophyllums lies in their machinery for trapping insects. Those who attended the Temple show last year saw something of it, if they could penetrate the crush around *B. barbigerum* on Sir Trevor Lawrence's stand. This tiny but amazing plant comes from Sierra Leone. The long yellow lip is attached to the column by the slenderest possible joint, so that it rocks without an instant's pause. At the tip is set a brush of silky hairs, which wave backwards and forwards with the precision of machinery. No wonder that the natives believe it a living thing. The purpose of these arrangements is to catch flies, [Pg which other species effect with equal ingenuity if less elaboration. Very pretty too are some of them, as *B. Lobbii*. Its clear, clean, orange-creamy hue is delightful to behold. The lip, so delicately balanced, quivers at every breath. If the slender stem be bent back, as by a fly alighting on the column, that quivering cap turns and hangs imminent; another tiny shake, as though the fly approached the nectary, and it falls plump, head over heels, like a shot, imprisoning the insect. Thus the flower is impregnated. If we wished to excite a thoughtful child's interest in botany — not regardless of the sense of beauty either — we should make an investment in *Bulbophyllum Lobbii*. *Bulbophyllum Dearei* also is pretty — golden ochre spotted red, with a wide dorsal sepal, very narrow petals flying behind, lower sepals broadly striped with red, and a yellow lip, upon a hinge, of course; but the gymnastic performances of this species are not so impressive as in most of its kin.

A new Bulbophyllum, *B. Godseffianum*, has lately been brought from the Philippines, contrived on the same principle, but even more charming. The flowers, two inches broad, have the colour of "old gold," with stripes of crimson on the petals, and the dorsal sepal shows membranes almost transparent, which have the effect of silver embroidery.

Until *B. Beccarii* was introduced, from Borneo, in 1867, the Grammatophyllums were regarded as monsters incomparable. Mr. Arthur Keyser, Resident Magistrate at Selangor, in the Straits Settlement, tells of one which he gathered on a Durian tree, seven feet two inches high, thirteen feet six inches across, bearing seven spikes of flower, the longest eight feet six inches — a weight which fifteen men could only just carry. Mr. F.W. Burbidge heard a tree fall in the jungle one night when he was four miles away, and on visiting the spot, he found, "right in the collar of the trunk, a Grammatophyllum big enough to fill a Pickford's van, just opening its golden-brown spotted flowers, on stout spikes two yards long." It is not to be hoped that we shall ever see monsters like these in Europe. The genus, indeed, is unruly. *G. speciosum* has been grown to six feet high, I believe, which is big enough to satisfy the modest amateur, especially when it develops leaves two feet long. The flowers are — that is, they ought to be — six inches in diameter, rich yellow, blotched with reddish purple. They have some giants at Kew now, of which fine things are expected. *G. Measureseanum*, named after Mr. Measures, a leading [Pg 172]amateur, is pale buff, speckled with chocolate, the ends of the sepals and petals charmingly tipped with the same hue. Within the last few months Mr. Sander has obtained *G. multiflorum* from the Philippines, which seems to be not only the most beautiful, but the easiest to cultivate of those yet introduced. Its flowers droop in a garland of pale green and yellow, splashed with brown, not loosely set, as is the rule, but scarcely half an inch apart. The effect is said to be lovely beyond description. We may hope to judge for ourselves in no long time, for Mr. Sander has presented a wondrous specimen to the Royal Gardens, Kew. This is assuredly the biggest orchid ever brought to Europe. Its snakey pseudo-bulbs measure nine feet, and the old flower spikes stood eighteen feet high. It will be found in the Victoria Regia house, growing strongly.

FOOTNOTES:

[6] *Vanda Lowii* is properly called *Renanthera Lowii*.

[7] *Vide* page 100.

[Pg 173]

THE LOST ORCHID.

Not a few orchids are "lost"—have been described that is, and named, even linger in some great collection, but, bearing no history, cannot now be found. Such, for instance, are *Cattleya Jongheana, Cymbidium Hookerianum, Cypripedium Fairianum*. But there is one to which the definite article might have been applied a very few days ago. This is *Cattleya labiata vera*. It was the first to bear the name of Cattleya, though not absolutely the first of that genus discovered. *C. Loddigesii* preceded it by a few years, but was called an Epidendrum. Curious it is to note how science has returned in this latter day to the views of a pre-scientific era. Professor Reichenbach was only restrained from abolishing the genus Cattleya, and merging all its species into Epidendrum, by regard for the weakness of human nature. *Cattleya labiata vera* was sent from Brazil to Dr. Lindley by Mr. W. Swainson, and reached Liverpool in 1818. So much is certain, for Lindley [Pg 174]makes the statement in his *Collectanea Bo-*

tanica. But legends and myths encircle that great event. It is commonly told in books that Sir W. Jackson Hooker, Regius Professor of Botany at Glasgow, begged Mr. Swainson—who was collecting specimens in natural history—to send him some lichens. He did so, and with the cases arrived a quantity of orchids which had been used to pack them. Less suitable material for "dunnage" could not be found, unless we suppose that it was thrust between the boxes to keep them steady. Paxton is the authority for this detail, which has its importance. The orchid arriving in such humble fashion proved to be *Cattleya labiata*; Lindley gave it that name—there was no need to add *vera* then. He established a new genus for it, and thus preserved for all time the memory of Mr. Cattley, a great horticulturist dwelling at Barnet. There was no ground in supposing the species rare. A few years afterwards, in fact, Mr. Gardner, travelling in pursuit of butterflies and birds, sent home quantities of a Cattleya which he found on the precipitous sides of the Pedro Bonita range, and also on the Gavea, which our sailors call "Topsail" Mountain, or "Lord Hood's Nose." These orchids passed as *C. labiata* for a while. Paxton congratulated himself and the world in [Pg 175]his *Flower Garden* that the stock was so greatly increased. Those were the coaching days, when botanists had not much opportunity for comparison. It is to be observed, also, that Gardner's Cattleya was the nearest relative of Swainson's;—it is known at present as *C. labiata Warneri*. The true species, however, has points unmistakable. Some of its kinsfolk show a double flower-sheath;—very, very rarely, under exceptional circumstances. But *Cattleya labiata vera* never fails, and an interesting question it is to resolve why this alone should be so carefully protected. One may cautiously surmise that its habitat is even damper than others'. In the next place, some plants have their leaves red underneath, others green, and the flower-sheath always corresponds; this peculiarity is shared by *C.l. Warneri* alone. Thirdly—and there is the grand distinction, the one which gives such extreme value to the species—it flowers in the late autumn, and thus fills a gap. Those who possess a plant may have Cattleyas in bloom the whole year round—and they alone. Accordingly, it makes a section by itself in the classification of *Reichenbachia*, as the single species that flowers from the current year's growth, after resting. Section II. contains the species that flower from the current [Pg 176]year's growth before resting. Section III., those that flower from last year's growth after resting. All these are many, but *C.l. vera* stands alone.

Cattleya labiata.
Reduced to One Sixth

We have no need to dwell upon the contest that arose at the introduction of *Cattleya Mossiæ* in 1840, which grew more and more bitter as others of the class came in, and has not yet ceased. It is enough to say that Lindley declined to recognize *C. Mossiæ* as a species, though he stood almost solitary against "the trade," backed by a host of enthusiastic amateurs. The great botanist declared that he could see nothing in the beautiful new Cattleya to distinguish it as a species from the one already named, *C. labiata*, except that most variable of characteristics, colour. Modes of growth and times of flowering do not concern science. The structure of the plants is identical, and to admit *C. Mossiæ* as a sub-species of the same was the utmost concession Lindley would make. This was in 1840. Fifteen years later came *C. Warscewiczi*, now called *gigas*; then, next year, *C. Trianæ*; *C. Dowiana* in 1866; *C. Mendellii* in 1870 — all *labiatas*, strictly speaking. At each arrival the controversy was renewed; it is not over yet. But Sir Joseph Hooker succeeded Lindley and Reichenbach succeeded Hooker as the supreme authority, and [Pg 177]each of them stood firm. There are, of course, many Cattleyas recognized as species, but Lindley's rule has been maintained. We may return to the lost orchid.

As time went on, and the merits of *C. labiata vera* were understood, the few specimens extant — proceeding from Mr. Swainson's importation — fetched larger and larger prices. Those merits, indeed, were conspicuous. Besides the season of flowering, this proved to be the strongest and most easily grown of Cattleyas. Its normal type was at least as charming as any, and it showed an extraordinary readiness to vary. Few, as has been said, were the plants in cultivation, but they gave three distinct varieties. Van Houtte shows us two in his admirable *Flore des Serres*; *C.l. candida*, from Syon House, pure white excepting the ochrous throat — which is invariable — and *C.l. picta*, deep red, from the collection of J.J. Blandy, Esq., Reading. The third was *C.l. Pescatorei*, white, with a deep red blotch upon the lip, formerly owned by Messrs. Rouget-Chauvier, of Paris, now by the Duc de Massa.

Under such circumstances the dealers began to stir in earnest. From the first, indeed, the more enterprising had made efforts to import a plant which, as they supposed, must be a common weed [Pg 178]at Rio, since men used it to "pack" boxes. But that this was an error they soon perceived. Taking the town as a centre, collectors pushed out on all sides. Probably there is not one of the large dealers, in England or the Continent, dead or living, who has not spent money — a large sum, too — in searching for *C. l. vera*. Probably, also, not one has lost by the speculation, though never a sign nor a hint, scarcely a rumour, of the thing sought rewarded them. For all secured new orchids, new bulbs — Eucharis in especial — Dipladenias, Bromeliaceæ, Calladiums, Marantas, Aristolochias, and what not. In this manner the lost orchid has done immense service to botany and to mankind. One may say that the hunt lasted seventy years, and led collectors to strike a path through almost every province of Brazil — almost, for there are still vast regions unexplored. A man might start, for example, at Para, and travel to Bogota, two thousand miles or so, with a stretch of six hundred miles on either hand which is untouched. It may well be asked what Mr. Swainson was doing, if alive, while his discovery thus agitated the world. Alive he was, in New Zealand, until the year 1855, but he offered no assistance. It is scarcely to be doubted that he had none to give. The orchids fell in his way [Pg 179]by accident — possibly collected in distant parts by some poor fellow who died at Rio. Swainson picked them up, and used them to stow his lichens.

Not least extraordinary, however, in this extraordinary tale is the fact that various bits of *C.l. vera* turned up during this time. Lord Home has a noble specimen at Bothwell Castle, which did not come from Swainson's consignment. His gardener told the story five years ago. "I am quite sure," he wrote, "that my nephew told me the small bit I had from him" — forty years before — "was off a newly-imported plant, and I understood it had been brought by one of Messrs. Horsfall's ships." Lord Fitzwilliam seems to have got one in the same way, from another ship. But the most astonishing case is recent. About seven years ago two plants made their appearance in the Zoological Gardens at Regent's Park — in the conservatory behind Mr. Bartlett's house. How they got there is an eternal mystery. Mr. Bartlett sold them for a large sum; but an equal sum offered him for any scrap of information showing how they came into his hands he was sorrowfully obliged to refuse — or, rather, found himself unable to earn. They certainly arrived in company with some monkeys; but when, from what district of South America, the closest search [Pg 180]of his papers failed to show. In 1885, Dr. Regel, Director of the Imperial Gardens at St. Petersburg, received a few plants. It may be worth while to name those gentlemen who recently possessed examples of *C.l. vera*, so far as our knowledge goes. They were Sir Trevor Lawrence, Lord Rothschild, Duke of Marlborough, Lord Home, Messrs. J. Chamberlain, T. Statten, J.J. Blandy, and G. Hardy, in England; in America, Mr. F.L. Ames, two, and Mr. H.H. Hunnewell; in France, Comte de Germiny, Duc de Massa, Baron Alphonse and Baron Adolf de Rothschild, M. Treyeran of Bordeaux. There were two, as is believed, in Italy.

And now the horticultural papers inform us that the lost orchid is found, by Mr. Sander of St. Albans. Assuredly he deserves his luck — if the result of twenty years' labour should be so described. It was about 1870, we believe, that Mr. Sander sent out Arnold, who passed five years in exploring Venezuela. He had made up his mind that the treasure must not be looked for in Brazil. Turning next to Colombia, in successive years, Chesterton, Bartholomeus, Kerbach, and the brothers Klaboch overran that country. Returning to Brazil, his collectors, Oversluys, Smith, Bestwood, went over every foot of the [Pg 181]ground which Swainson seems, by his books, to have traversed. At the same time Clarke followed Gardner's track through the Pedro Bonita and Topsail Mountains. Then Osmers traced the whole coast-line of the Brazils from north to south, employing five years in the work. Finally, Digance undertook the search, and died this year. To these men we owe grand discoveries beyond counting. To name but the grandest, Arnold found *Cattleya Percevaliana*; from Colombia were brought *Odont. vex. rubellum*, *Bollea cœlestis*, *Pescatorea Klabochorum*; Smith sent *Cattleya O'Brieniana*; Clarke the dwarf Cattleyas, *pumila* and *præstans*; Lawrenceson *Cattleya Schroederæ*; Chesterton *Cattleya Sanderiana*; Digance *Cattleya Diganceana*, which received a Botanical certificate from the Royal Horticultural Society on September 8th, 1890. But they heard not a whisper of the lost orchid.

In 1889 a collector employed by M. Moreau, of Paris, to explore Central and North Brazil in search of insects, sent home fifty plants—for M. Moreau is an enthusiast in orchidology also. He had no object in keeping the secret of its habitat, and when Mr. Sander, chancing to call, recognized the treasure so long lost, he gave every assistance. Meanwhile, the International Horticultural Society of Brussels [Pg 182]had secured a quantity, but they regarded it as new, and gave it the name of *Catt. Warocqueana*; in which error they persisted until Messrs. Sander flooded the market.

[Pg 183]

AN ORCHID FARM.

My articles brought upon me a flood of questions almost as embarrassing as flattering to a busy journalist. The burden of them was curiously like. Three ladies or gentlemen in four wrote thus: "I love orchids. I had not the least suspicion that they may be cultivated so easily and so cheaply. I am going to begin. Will you please inform me"—here diversity set in with a vengeance! From temperature to flower-pots, from the selection of species to the selection of peat, from the architecture of a greenhouse to the capabilities of window-gardening, with excursions between, my advice was solicited. I replied as best I could. It must be feared, however, that the most careful questioning and the most elaborate replies by post will not furnish that ground-work of knowledge, the ABC of the science, which is needed by a person utterly unskilled; nor will he find it readily in the hand-books. Written by men familiar with the alphabet of orchidology [Pg 184]from their youth up, though they seem to begin at the beginning, ignorant enthusiasts who study them find woeful gaps. It is little I can do in this matter; yet, believing that the culture of these plants will be as general shortly as the culture of pelargoniums under glass—and firmly convinced that he who hastens that day is a real benefactor to his kind—I am most anxious to do what lies in my power. Considering the means by which this end may be won, it appears necessary above all to avoid boring the student. He should be led to feel how charming is the business in hand even while engaged with prosaic details; and it seems to me, after some thought, that the sketch of a grand orchid nursery will best serve our purpose for the moment. There I can show at once processes and results, passing at a step as it were from the granary into the harvest-field, from the workshop to the finished and glorious production.

"An orchid farm" is no extravagant description of the establishment at St. Albans. There alone in Europe, so far as I know, three acres of ground are occupied by orchids exclusively. It is possible that larger houses might be found—everything is possible; but such are devoted more or less to a variety of plants, and the departments are not all [Pg 185]gathered beneath one roof. I confess, for my own part, a hatred of references. They interrupt the writer, and they distract the reader. At the place I have chosen to illustrate our theme, one has but to cross a corridor from any of the working quarters to reach the showroom. We may start upon our critical survey from the very dwelling-house. Pundits of agricultural science explore the sheds, I believe, the barns, stables, machine-rooms, and so forth, before inspecting the crops. We may follow the same course, but our road offers an unusual distraction.

It passes from the farmer's hall beneath a high glazed arch. Some thirty feet beyond, the path is stopped by a wall of tufa and stalactite which rises to the lofty roof, and compels the traveller to turn right or left. Water pours down it and falls trickling into a narrow pool beneath. Its rough front is studded with orchids from crest to base. Cœlogenes have lost those pendant wreaths of bloom which lately tipped the rock as with snow. But there are Cymbidiums arching long sprays of green and chocolate; thickets of Dendrobe set with flowers beyond counting—ivory and rose and purple and orange; scarlet Anthuriums: huge clumps of Phajus and evergreen Calanthe, with a score of spikes rising from their broad leaves; [Pg 186] Cypripediums of quaint form and striking half-tones of colour; Oncidiums which droop their slender garlands a yard long, golden yellow and spotted, purple and white—a hundred tints. The crown of the rock bristles all along with Cattleyas, a dark-green glossy little wood against the sky. The *Trianæs* are almost over, but here and there a belated beauty pushes through, white or rosy, with a lip of crimson velvet. *Mossiæs* have replaced them generally, and from beds three feet in diameter their great blooms start by the score, in every shade of pink and crimson and rosy purple. There is *Lælia elegans*, exterminated in its native home, of such bulk and such luxuriance of growth that the islanders left forlorn might almost find consolation in regarding it here. Over all, climbing up the spandrils of the roof in full blaze of sunshine, is *Vanda teres*, round as a pencil both leaves and stalk, which will drape those bare iron rods presently with crimson and pink and gold. [8] The way to our farmyard is not like others. It traverses a corner of fairyland.

We find a door masked by such a rock as that faintly and vaguely pictured, which opens on a broad corridor. Through all its length, four hundred feet, it is ceilinged with baskets of [Pg 187] Mexican orchid, as close as they will fit. Upon the left hand lie a series of glass structures; upon the right, below the level of the corridor, the workshops; at the end—why, to be frank, the end is blocked by a ponderous screen of matting just now. But this dingy barrier is significant of a work in hand which will not be the least curious nor the least charming of the strange sights here. The farmer has already a "siding" of course, for the removal of his produce; he finds it necessary to have a station of his own also for the convenience of clients. Beyond the screen at present lies an area of mud and ruin, traversed by broken walls and rows of hot-water piping swathed in felt to exclude the chill air. A few weeks since, this little wilderness was covered with glass, but the ends of the long "houses" have been cut off to make room for a structure into which visitors will step direct from the train. The platform is already finished, neat and trim; so are the vast boilers and furnaces, newly rebuilt, which would drive a cotton factory.

A busy scene that is which we survey, looking down through openings in the wall of the corridor. Here is the composing-room, where that magnificent record of orchidology in three languages, the "Reichenbachia," slowly advances from year to [Pg 188]year. There is the printing-room, with no steam presses or labour-saving machinery, but the most skilful craftsmen to be found, the finest paper, the most deliberate and costly processes, to rival the great works of the past in illustrating modern science. These departments, however, we need not visit, nor the chambers, lower still, where mechanical offices are performed.

The "Importing Room" first demands notice. Here cases are received by fifties and hundreds, week by week, from every quarter of the orchid world, unpacked, and their contents stored until space is made for them up above. It is a long apartment, broad and low, with tables against the wall and down the middle, heaped with things which to the uninitiated seem, for the most part, dry sticks and dead bulbs. Orchids everywhere! They hang in dense bunches from the roof. They lie a foot thick upon every board, and two feet thick below. They are suspended on the walls. Men pass incessantly along the gangways, carrying a load that would fill a barrow. And all the while fresh stores are accumulating under the hands of that little group in the middle, bent and busy at cases just arrived. They belong to a lot of eighty that came in from Burmah last night—and while we look on, a boy brings a telegram [Pg 189]announcing fifty more from Mexico, that will reach Waterloo at 2.30 p.m. Great is the wrath and great the anxiety at this news, for some one has blundered; the warning should have been despatched three hours before. Orchids must not arrive at unknown stations unless there be somebody of discretion and experience to meet them, and the next train does not leave St. Albans until 2.44 p.m. Dreadful is the sense of responsibility, alarming the suggestions of disaster, that arise from this incident.

The Burmese cases in hand just now are filled with Dendrobiums, *crassinode* and *Wardianum*, stowed in layers as close as possible, with *D. Falconerii* for packing material. A royal way of doing things indeed to substitute an orchid of value for shavings or moss, but mighty convenient and profitable. For that packing will be sent to the auction-rooms presently, and will be sold for no small proportion of the sum which its more delicate charge attains. We remark that the experienced persons who remove these precious sticks, layer by layer, perform their office gingerly. There is not much danger or unpleasantness in unpacking Dendrobes, compared with other genera, but ship-rats spring out occasionally and give an ugly bite; scorpions and [Pg 190]centipedes have been known to harbour in the close roots of *D. Falconerii*; stinging ants are by no means improbable, nor huge spiders; while cockroaches of giant size, which should be killed, may be looked for with certainty. But men learn a habit of caution by experience of cargoes much more perilous. In those masses of *Arundina bambusæfolia* beneath the table yonder doubtless there are centipedes lurking, perhaps even scorpions, which have escaped the first inspection. Happily, these pests are dull, half-stupefied with the cold, when discovered, and no man here has been stung, circumspect as they are; but ants arrive as alert and as vicious as in their native realm. Distinctly they are no joke. To handle a consignment of *Epidendrum bicornutum* demands some nerve. A very ugly species loves its hollow bulbs, which, when disturbed, shoots out with lightning swiftness and nips the arm or hand so quickly that it can seldom be avoided. But the most awkward cases to deal with are those which contain *Schomburghkia tibicinis*. This superb orchid is so difficult to bloom that very few will attempt it; I have seen its flower but twice. Packers strongly approve the reluctance of the public to buy, since it restricts importation. The foreman has been laid up again and again. But they [Pg 191]find pleasing curiosities also, tropic beetles, and insects, and cocoons. Dendrobiums in especial are favoured by moths; *D. Wardianum* is loaded with their webs, empty as a rule. Hitherto the men have preserved no chrysalids, but at this moment they have a few, of unknown species.

The farmer gets strange bits of advice sometimes, and strange offers of assistance. Talking of insects reminds him of a letter received last week. Here it is:—

Sirs,—I have heard that you are large growers of orchids; am I right in supposing that in their growth or production you are much troubled with some insect or caterpillar which retards or hinders their arrival at maturity, and that these insects or caterpillars can be destroyed by small snakes? I have tracts of land under my occupation, and if these small snakes can be of use in your culture of orchids you might write, as I could get you some on knowing what these might be worth to you.

Yours truly
— —

Thence we mount to the potting-rooms, where a dozen skilled workmen try to keep pace with the growth of the imported plants; taking up, day by day, those which thrust out roots so fast that postponement is injurious. The broad middle tables are heaped with peat and moss and leaf-mould and white sand. At counters on either side unskilled labourers are sifting and mixing, while boys come and go, laden with pots and baskets of [Pg 192]teak-wood and crocks and charcoal. These things are piled in heaps against the walls; they are stacked on frames overhead; they fill the semi-subterranean chambers of which we get a glimpse in passing. Our farm resembles a factory in this department.

Ascending to the upper earth again, and crossing the corridor, we may visit number one of those glass-houses opposite. I cannot imagine, much more describe, how that spectacle would strike one to whom it was wholly unfamiliar. These buildings—there are twelve of them, side by side—measure one hundred and eighty feet in length, and the narrowest has thirty-two feet breadth. This which we enter is devoted to *Odontoglossum crispum*, with a few *Masdevallias*. There were twenty-two thousand pots in it the other day; several thousand have been sold, several thousand have been brought in, and the number at this moment cannot be computed. Our farmer has no time for speculative arithmetic; he deals in produce wholesale. Telegraph an order for a thousand *crispums* and you cause no stir in the establishment. You take it for granted that a large dealer only could propose such a transaction. But it does not follow at all. Nobody would credit, unless he had talked with one of the great farmers, on what enormous scale [Pg 193]orchids are cultivated up and down by private persons. Our friend has a client who keeps his stock of *O. crispum* alone at ten thousand; but others, less methodical, may have more.

Opposite the door is a high staging, mounted by steps, with a gangway down the middle and shelves descending on either hand. Those shelves are crowded with fine plants of the glorious *O. crispum*, each bearing one or two spikes of flower, which trail down, interlace, arch upward. Not all are in bloom; that amazing sight may be witnessed for a month to come—for two months, with such small traces of decay as the casual visitor would not notice. So long and dense are the wreaths, so broad the flowers, that the structure seems to be festooned from top to bottom with snowy garlands. But there is more. Overhead hang rows of baskets, lessening in perspective, with pendent sprays of bloom. And broad tables which edge the walls beneath that staging display some thousands still, smaller but not less beautiful. A sight which words could not portray. I yield in despair.

The tillage of the farm is our business, and there are many points here which the amateur should note. Observe the bricks beneath your feet. They have a hollow pattern which retains [Pg 194]the water, though your boots keep dry. Each side of the pathway lie shallow troughs, always full. Beneath that staging mentioned is a bed of leaves, interrupted by a tank here, by a group of ferns there, vividly green. Slender iron pipes run through the house from end to end, so perforated that on turning a tap they soak these beds, fill the little troughs and hollow bricks, play in all directions down below, but never touch a plant. Under such constant drenching the leaf-beds decay, throwing up those gases and vapours in which the orchid delights at home. Thus the amateur should arrange his greenhouse, so far as he may. But I would not have it understood that these elaborate contrivances are essential. If you would beat Nature, as here, making invariably such bulbs and flowers as she produces only under rare conditions, you must follow this system. But orchids are not exacting.

The house opens, at its further end, in a magnificent structure designed especially to exhibit plants of warm species in bloom. It is three hundred feet long, twenty-six wide, eighteen high—the piping laid end to end, would measure as nearly as possible one mile: we see a practical illustration of the resources of the establishment, when it is expected to furnish such a show. Here [Pg 195]are stored the huge specimens of *Cymbidium Lowianum*, nine of which astounded the good people of Berlin with a display of one hundred and fifty flower spikes, all open at once. We observe at least a score as well furnished, and hundreds which a royal gardener would survey with pride. They rise one above another in a great bank, crowned and brightened by garlands of pale green and chocolate. Other Cymbidiums are here, but not the beautiful *C. eburneum*. Its large white flowers, erect on a short spike, not drooping like these, will be found in a cool house—smelt with delight before they are found.

Further on we have a bank of Dendrobiums, so densely clothed in bloom that the leaves are unnoticed. Lovely beyond all to my taste, if, indeed, one may make a comparison, is *D. luteolum*, with flowers of palest, tenderest primrose, rarely seen unhappily, for it will not reconcile itself to our treatment. Then again a bank of Cattleyas, of

Vandas, of miscellaneous genera. The pathway is hedged on one side with *Begonia coralina*, an unimproved species too straggling of growth and too small of flower to be worth its room under ordinary conditions; but a glorious thing here, climbing to the roof, festooned at every season of the year with countless rosy sprays.

[Pg 196]Beyond this show-house lie the small structures devoted to "hybridization," but I deal with them in another chapter. Here also are the Phalœnopsis, the very hot Vandas, Bolleas, Pescatoreas, Anæctochili, and such dainty but capricious beauties.

We enter the second of the range of greenhouses, also devoted to Odontoglossums, Masdevallias, and "cool" genera, as crowded as the last; pass down it to the corridor, and return through number three, which is occupied by Cattleyas and such. There is a lofty mass of rock in front, with a pool below, and a pleasant sound of splashing water. Many orchids of the largest size are planted out here — Cypripedium, Cattleya, Sobralia, Phajus, Lœlia, Zygopetalum, and a hundred more, "specimens," as the phrase runs — that is to say, they have ten, twenty, fifty, flower spikes. I attempt no more descriptions; to one who knows, the plain statement of fact is enough, one who does not is unable to conceive that sight by the aid of words. But the Sobralias demand attention. They stand here in clumps two feet thick, bearing a wilderness of loveliest bloom — like Irises magnified and glorified by heavenly enchantment. Nature designed a practical joke perhaps when she granted these noble flowers but one day's existence each, while dingy Epidendrums last six months, or nine. I imagine that for stateliness and delicacy combined there are no plants that excel the Sobralia. At any single point they may be surpassed — among orchids, be it understood, by nothing else in Nature's realm — but their magnificence and grace together cannot be outshone.

I must not dwell upon the marvels here, in front, on either side, and above — a hint is enough. There are baskets of *Lœlia anceps* three feet across, lifted bodily from the tree in their native forest where they had grown perhaps for centuries. One of them — the white variety, too, which æsthetic infidels might adore, though they believed in nothing — opened a hundred spikes at Christmas time; we do not concern ourselves with minute reckonings here. [Pg 197] But an enthusiastic novice counted the flowers blooming one day on that huge mass of *Lælia albida* yonder, and they numbered two hundred and eleven—unless, as some say, this was the quantity of "spikes," in which case one must have to multiply by two or three. Such incidents maybe taken for granted at the farm.

Loelianceps Schroederiana.
Reduced to One Sixth

But we must not pass a new orchid, quite distinct and supremely beautiful, for which Professor Reichenbach has not yet found a name sufficiently appreciative. Only eight pieces were discovered, whence we must suspect that it is very rare at [Pg 198]home; I do not know where the home is, and I should not tell if I did. Such information is more valuable than the surest tip for the Derby, or most secrets of State. This new orchid is a Cyrrhopetalun, of very small size, but, like so many others, its flower is bigger than itself. The spike inclines almost at a right angle, and the pendent half is hung with golden bells, nearly two inches in length. Beneath it stands the very rare scarlet Utricularia, growing in the axils of its native Vriesia, as in a cup always full; but as yet the flower has been seen in Europe only by the eyes of faith. It may be news to some that Utricularias do not belong to the orchid family—have, in fact, not the slightest kinship, though associated with it by growers to the degree that Mr. Sander admits them to his farm. A little story hangs to the exquisite *U. Campbelli*. All importers are haunted by the spectral image of *Cattleya labiata*, which, in its true form, had been brought to Europe only once, seventy years ago, when this book was written. Some time since, Mr. Sander was looking through the drawings of Sir Robert Schomburgk, in the British Museum, among which is a most eccentric Cattleya named—for reasons beyond comprehension—a variety of *C. Mossiæ*. He jumped at the conclusion that this must be the long-lost *C. labiata*. [Pg 199] So strong indeed was his confidence that he despatched a man post-haste over the Atlantic to explore the Roraima mountain; and, further, gave him strict injunctions to collect nothing but this precious species. For eight months the traveller wandered up and down among the Indians, searching forest and glade, the wooded banks of streams, the rocks and clefts, but he found neither *C. labiata* nor that curious plant which Sir Robert Schomburgk described. Upon the other hand, he came across the lovely *Utricularia Campbelli*, and in defiance of instructions brought it down. But very few reached England alive. For six weeks they travelled on men's backs, from their mountain home to the River Essequibo; thence, six weeks in canoe to

Georgetown, with twenty portages; and, so aboard ship. The single chance of success lies in bringing them down, undisturbed, in the great clumps of moss which are their habitat, as is the Vriesia of other species.

I will allow myself a very short digression here. It may seem unaccountable that a plant of large growth, distinct flower, and characteristic appearance, should elude the eye of persons trained to such pursuits, and encouraged to spend money on the slightest prospect of success, for half a century and more. But if we recall the circumstances it [Pg 200]ceases to astonish. I myself spent many months in the forests of Borneo, Central America, and the West African coast. After that experience I scarcely understand how such a quest, for a given object, can ever be successful unless by mere fortune. To look for a needle in a bottle of hay is a promising enterprise compared with the search for an orchid clinging to some branch high up in that green world of leaves. As a matter of fact, collectors seldom discover what they are specially charged to seek, if the district be untravelled—the natives, therefore, untrained to grasp and assist their purpose. This remark does not apply to orchids alone; not by any means. Few besides the scientific, probably, are aware that the common *Eucharis amasonica* has been found only once; that is to say, but one consignment has ever been received in Europe, from which all our millions in cultivation have descended. Where it exists in the native state is unknown, but assuredly this ignorance is nobody's fault. For a generation at least skilled explorers have been hunting. Mr. Sander has had his turn, and has enjoyed the satisfaction of discovering species closely allied, as *Eucharis Mastersii* and *Eucharis Sanderiana*; but the old-fashioned bulb is still to seek.

In this third greenhouse is a large importation of [Pg 201] *Cattleya Trianæ*, which arrived so late last year that their sheaths have opened contemporaneously with *C. Mossiæ*. I should fear to hazard a guess how many thousand flowers of each are blooming now. As the Odontoglossums cover their stage with snow wreaths, so this is decked with upright plumes of *Cattleya Trianæ*, white and rose and purple in endless variety of tint, with many a streak of other hue between.

Suddenly our guide becomes excited, staring at a basket overhead beyond reach. It contains a smooth-looking object, very green and fat, which must surely be good to eat—but this observation is alike irrelevant and disrespectful. Why, yes! Beyond all possibility of doubt that is a spike issuing from the axil of its fleshy leaf! Three inches long it is already, thick as a pencil, with a big knob of bud at the tip. Such pleasing surprises befall the orchidacean! This plant came from Borneo so many years ago that the record is lost; but the oldest servant of the farm remembers it, as a poor cripple, hanging between life and death, season after season. Cheerful as interesting is the discussion that arises. More like a Vanda than anything else, the authorities resolve, but not a Vanda! Commending it to the special care of those responsible, we pass on.

[Pg 202]Here is the largest mass of Catasetum ever found, or even rumoured, lying in ponderous bulk upon the stage, much as it lay in a Guatemalan forest. It is engaged in the process of "plumping up." Orchids shrivel in their long journey, and it is the importer's first care to renew that smooth and wholesome rotundity which indicates a conscience untroubled, a good digestion, and an assurance of capacity to fulfil any reasonable demand. Beneath the staging you may see myriads of withered sticks, clumps of shrunken and furrowed bulbs by the thousand, hung above those leaf-beds mentioned; they are "plumping" in the damp shade. The larger pile of Catasetum—there are two—may be four feet long, three wide, and eighteen inches thick; how many hundreds of flowers it will bear passes computation. I remarked that when broken up into handsome pots it would fill a greenhouse of respectable dimensions; but it appears that there is not the least intention of dividing it. The farmer has several clients who will snap at this natural curiosity, when, in due time, it is put on the market.

At the far end of the house stands another piece of rockwork, another little cascade, and more marvels than I can touch upon. In fact, there are [Pg 203]several which would demand all the space at my disposition, but, happily, one reigns supreme. This is a *Cattleya Mossiæ*, the pendant of the Catasetum, by very far the largest orchid of any kind that was ever brought to Europe. For some years Mr. Sander, so to speak, hovered round it, employing his shrewdest and most diplomatic agents. For this was not a forest specimen. It grew upon a high tree beside an Indian's hut, near Caraccas, and belonged to him as absolutely as the fruit in his compound. His great-grandfather, indeed, had "planted" it, so he declared, but this is highly improbable. The giant has embraced two stems of the tree, and covers them both so thickly that the bare ends of wood at top alone betray its secret; for it was sawn off, of course, above and below. I took the dimensions as accurately as may be, with an object so irregular and prickly. It measures—the solid bulk of it, leaves not counted—as nearly as possible five feet in height and four thick—one plant, observe, pulsating through its thousand limbs from one heart; at least, I mark no spot where the circulation has been checked by accident or disease, and the pseudo-bulbs beyond have been obliged to start an independent existence.

In speaking of *Lælia elegans*, I said that those [Pg 204] Brazilian islanders who have lost it might find solace could they see its happiness in exile. The gentle reader thought this an extravagant figure of speech, no doubt, but it is not wholly fanciful. Indians of Tropical America cherish a fine orchid to the degree that in many cases no sum, and no offer of valuables, will tempt them to part with it. Ownership is distinctly recognized when the specimen grows near a village. The root of this feeling, whether superstition or taste, sense of beauty, rivalry in magnificence of church displays, I have not been able to trace. It runs very strong in Costa Rica, where the influence of the aborigines is scarcely perceptible, and there, at least, the latter motive is sufficient explanation. Glorious beyond all our fancy can conceive, must be the show in those lonely forest churches, which no European visits save the "collector," on a feast day. Mr. Roezl, whose name is so familiar to botanists, left a description of the scene that time he first beheld the Flor de Majo. The church was hung with garlands of it, he says, and such emotions seized him at the view that he choked. The statement is quite credible. Those who see that wonder now, prepared for its transcendent glory, find no words to express their feeling: imagine an enthusiast beholding it for the first time, unwarned, [Pg 205]unsuspecting that earth can show such a sample of the flowers that bloomed in Eden! And not a single branch, but garlands of it! Mr. Roezl proceeds to speak of bouquets of *Masdevallia Harryana* three feet across, and so forth. The natives showed him "gardens" devoted to this species, for the ornament of their church; it was not cultivated, of course, but evidently planted. They were acres in extent.

The Indian to whom this *Cattleya Mossiæ* belonged refused to part with it at any price for years; he was overcome by a rifle of peculiar fascination, added to the previous offers. A magic-lantern has very great influence in such cases, and the collector provides himself with one or more nowadays as part of his outfit. Under that charm, with 47*l*. in cash, Mr. Sander secured his first *C. Mossiæ alba*, but it has failed hitherto in another instance, though backed by 100*l*., in "trade" or dollars, at the Indian's option.

Thence we pass to a wide and lofty house which was designed for growing *Victoria Regia* and other tropic water-lilies. It fulfilled its purpose for a time, and I never beheld those plants under circumstances so well fitted to display their beauty. But they generate a small black fly in myriads beyond belief, and so the culture of *Nymphæa* was dropped. A [Pg 206]few remain, in manageable quantities, just enough to adorn the tank with blue and rosy stars; but it is arched over now with baskets as thick as they will hang — Dendrobium, Cœlogene, Oncidium, Spathoglottis, and those species which love to dwell in the neighbourhood of steaming water. My vocabulary is used up by this time. The wonders here must go unchronicled.

We have viewed but four houses out of twelve, a most cursory glance at that! The next also is intermediate, filled with Cattleyas, warm Oncidiums, Lycastes, Cypripediums — the inventory of names alone would occupy all my space remaining. At every step I mark some object worth a note, something that recalls, or suggests, or demands a word. But we must get along. The sixth house is cool again — Odontoglossums and such; the seventh is given to Dendrobes. But facing us as we enter stands a *Lycaste Skinneri*, which illustrates in a manner almost startling the infinite variety of the orchid. I positively dislike this species, obtrusive, pretentious, vague in colour, and stiff in form. But what a royal glorification of it we have here! — what exquisite veining and edging of purple or rose; what a velvet lip of crimson darkening to claret! It is merely a sport of Nature, but she allows herself such glorious freaks [Pg 207]in no other realm of her domain. And here is a new Brassia just named by the pontiff of orchidology, Professor Reichenbach. Those who know the tribe of Brassias will understand why I make no effort to describe it. This wonderful thing is yet more "all over the shop" than its kindred. Its dorsal sepal measures three inches in length, its "tail," five inches, with an enormous lip between. They term it the Squid Flower, or Octopus, in Mexico; and a good name too. But in place of the rather weakly colouring habitual it has a grand decision of character, though the tones are like—pale yellow and greenish; its raised spots, red and deep green, are distinct as points of velvet upon muslin.

In the eighth house we return to Odontoglossums and cool genera. Here are a number of Hybrids of the "natural class," upon which I should have a good deal to say if inexorable fate permitted; "natural hybrids" are plants which seem species, but, upon thoughtful examination and study, are suspected to be the offspring of kindred and neighbours. Interesting questions arise in surveying fine specimens side by side, in flower, all attributed to a cross between *Odontoglossum Lindleyanum* and *Odontoglossum crispum Alexandræ*, and all quite different. But we must [Pg 208]get on to the ninth house, from which the tenth branches.

Here is the stove, and twilight reigns over that portion where a variety of super-tropic genera are "plumping up," making roots, and generally reconciling themselves to a new start in life. Such dainty, delicate souls may well object to the apprenticeship. It must seem very degrading to find themselves laid out upon a bed of cinders and moss, hung up by the heels above it, and even planted therein; but if they have as much good sense as some believe, they may be aware that it is all for their good. At the end, in full sunshine, stands a little copse of *Vanda teres*, set as closely as their stiff branches will allow. Still we must get on. There are bits of wood hanging here so rotten that they scarcely hold together; faintest dots of green upon them assure the experienced that presently they will be draped with pendant leaves, and presently again, we hope, with blue and white and scarlet flowers of Utricularia.

From the stove opens a very long, narrow house, where cool genera are "plumping," laid out on moss and potsherds; many of them have burst into strong growth. Pleiones are flowering freely as they lie. This farmer's crops come to harvest faster than he can attend to them. Things beauti [Pg 209]ful and rare and costly are measured here by the yard—so many feet of this piled up on the stage, so many of the other, from all quarters of the world, waiting the leisure of these busy agriculturists. Nor can we spare them more than a glance. The next house is filled with Odontoglossums, planted out like "bedding stuff" in a nursery, awaiting their turn to be potted. They make a carpet so close, so green, that flowers are not required to charm the eye as it surveys the long perspective. The rest are occupied just now with cargoes of imported plants.

My pages are filled—to what poor purpose, seeing how they might have been used for such a theme, no one could be so conscious as I.

FOOTNOTES:

[8] I was too sanguine. *Vanda teres* refused to thrive.

[Pg 210]

ORCHIDS AND HYBRIDIZING.

In the very first place, I declare that this is no scientific chapter. It is addressed to the thousands of men and women in the realm who tend a little group of orchids lovingly, and mark the wonders of their structure with as much bewilderment as interest. They read of hybridization, they see the result in costly specimens, they get books, they study papers on the subject. But the deeper their research commonly, the more they become convinced that these mysteries lie beyond their attainment. I am not aware of any treatise which makes a serious effort to teach the uninitiated. Putting technical expressions on one side—though that obstacle is grave enough—every one of those which have come under my notice takes the mechanical preliminaries for granted. All are written by experts for experts. My purpose is contrary. I wish to show how it is done so clearly that a child or the dullest gardener may be able to perform the operations—so very easy when you know how to set to work.

Cypripedium (hybridum) Pollettianum.
Reduced to One Sixth

[Pg 211]After a single lesson, in the genus *Cypripedium* alone, a young lady of my household amused herself by concerting the most incredible alliances—*Dendrobium* with *Odontoglossum, Epidendrum* with *Oncidium, Oncidium* with *Odontoglossum*, and so forth. It is unnecessary to tell the experienced that in every case the seed vessel swelled; that matter will be referred to presently. I mention the incident only to show how simple are these processes if the key be grasped.

Amateur hybridizers of an audacious class are wanted because, hitherto, operators have kept so much to the beaten paths. The names of Veitch and Dominy and Seden will endure when those of great *savants* are forgotten; but business men have been obliged to concentrate their zeal upon experiments that pay. Fantastic crosses mean, in all probability, a waste of time, space, and labour; in fact, it is not until recent years that such attempts could be regarded as serious. So much the more creditable, therefore, are Messrs. Veitch's exertions in that line.

But it seems likely to me that when hybridizing becomes a common pursuit with those who grow orchids—and the time approaches fast—a very strange revolution may follow. It will appear, as I think, that the enormous list of pure species— [Pg 212]even genera—recognized at this date may be thinned in a surprising fashion. I believe—timidly, as becomes the unscientific—that many distinctions which anatomy recognizes at present as essential to a true species will be proved, in the future, to result from promiscuous hybridization through æons of time. "Proved," perhaps, is the word too strong, since human life is short; but such a mass of evidence will be collected that reasonable men can entertain no doubt. Of course the species will be retained, but we shall know it to be a hybrid—the offspring, perhaps, of hybrids innumerable.

I incline more and more to think that even genera may be disturbed in a surprising fashion, and I know that some great authorities agree with me outright, though they are unprepared to commit themselves at present. A very few years ago this suggestion would have been absurd, in the sense that it wanted facts in support. As our ancestors made it an article of faith that to fertilize an orchid was impossible for man, so we imagined until lately that genera would not mingle. But this belief grows unsteady. Though bi-generic crosses have not been much favoured, as offering little prospect of success, such results have been obtained already that the field of speculation [Pg 213]lies open to irresponsible persons like myself. When Cattleya has been allied with Sophronitis, Sophronitis with Epidendrum, Odontoglossum with Zygopetalum, Cœlogene with Calanthe, one may credit almost anything. What should be stated on the other side will appear presently.

How many hybrids have we now, established, and passing from hand to hand as freely as natural species? There is no convenient record; but in the trade list of a French dealer those he is prepared to supply are set apart with Gallic precision. They number 416; but imagination and commercial enterprise are not less characteristic of the Gaul than precision.

In the excellent "Manual" of Messrs. Veitch, which has supplied me with a mass of details, I find ten hybrid Calanthes; thirteen hybrid Cattleyas, and fifteen Lœlias, besides sixteen "natural hybrids"—species thus classed upon internal evidence—and the wondrous Sophro-Cattleya, bi-generic; fourteen Dendrobiums and one natural; eighty-seven Cypripediums—but as for the number in existence, it is so great, and it increases so fast, that Messrs. Veitch have lost count; Phajus one, but several from alliance with Calanthe; Chysis two; Epidendrum one; Miltonia one, and two [Pg 214]natural; Masdevallia ten, and two natural; and so on. And it must be borne in mind that these amazing results have been effected in one generation. Dean Herbert's achievements eighty years ago were not chronicled, and it is certain that none of the results survive. Mr. Sander of St. Albans preserves an interesting relic, the only one as yet connected with the science of orchidology. This is *Cattleya hybrida*, the first of that genus raised by Dominy, manager to Messrs. Veitch, at the suggestion of Mr. Harris of Exeter, to the stupefaction of our grandfathers. Mr. Harris will ever be remembered as the gentleman who showed Mr. Veitch's agent how orchids are fertilized, and started him on his career. This plant was lost for years, but Mr. Sander found it by chance in the collection of Dr. Janisch at Hamburg, and he keeps it as a curiosity, for in itself the object has no value. But this is a digression.

Dominy's earliest success, actually the very first of garden hybrids to flower—in 1856—was *Calanthe Dominii*, offspring of *C. Masuca* × *C. furcata*;—be it here remarked that the name of the mother, or seed parent, always stands first. Another interest attaches to *C. Dominii*. Both its parents belong to the *Veratræfolia* section of Calanthe, [Pg 215]the terrestrial species, and no other hybrid has yet been raised among them. We have here one of the numberless mysteries disclosed by hybridization. The epiphytal Calanthes, represented by *C. vestita*, will not cross with the terrestrial, represented by *C. veratræfolia*, nor will the mules of either. We may "give this up" and proceed. In 1859 flowered *C. Veitchii*, from *C. rosea*, still called, as a rule, *Limatodes rosea*, × *C. vestita*. No orchid is so common as this, and none more simply beautiful. But although the success was so striking, and the way to it so easy, twenty years passed before even Messrs. Veitch raised another hybrid Calanthe. In 1878 Seden flowered *C. Sedeni* from *C. Veitchii* × *C. vestita*. Others entered the field then, especially Sir Trevor Lawrence, Mr. Cookson, and Mr. Charles Winn. But the genus is small, and they mostly chose the same families, often giving new names to the progeny, in ignorance of each other's labour.

The mystery I have alluded to recurs again and again. Large groups of species refuse to inter-marry with their nearest kindred, even plants which seem identical in the botanist's point of view. There is good ground for hoping, however, that longer and broader experience will annihilate some at least of the axioms current in this matter. [Pg 216] Thus, it is repeated and published in the very latest editions of standard works that South American Cattleyas, which will breed, not only among themselves, but also with the Brazilian Lœlias, decline an alliance with their Mexican kindred. But Baron Schroeder possesses a hybrid of such typical parentage as *Catt. citrina*, Mexican, and *Catt. intermedia*, Brazilian. It was raised by Miss Harris, of Lamberhurst, Kent, one single plant only; and it has flowered several times. Messrs. Sander have crossed *Catt. guttata Leopoldii*, Brazil, with *Catt. Dowiana*, Costa Rica, giving *Catt. Chamberliana*; *Lœlia crispa*, Brazil, with the same, giving *Lœlio-Cattleya*

Pallas; Catt. citrina, Mexico, with *Catt. intermedia*, Brazil, giving *Catt. citrina intermedia* (Lamberhurst hybrid); *Lælia flava*, Brazil, with *Catt. Skinneri*, Costa Rica, giving *Lælio-Catt. Marriottiana; Lælia pumila*, Brazil, with *Catt. Dowiana*, Costa Rica, giving *Lælio-Catt. Normanii; Lælia Digbyana*, Central America, with *Catt. Mossiæ*, Venezuela, giving *Lælio-Catt. Digbyana-Mossiæ; Catt. Mossiæ*, Venezuela, with *Lælia cinnabarina*, Brazil, giving *Lælio-Catt. Phoebe*. Not yet flowered and unnamed, raised in the Nursery, are *Catt. citrina*, Mexico, with *Lælia purpurata*, Brazil; *Catt. Harrisoniæ*, Brazil, with *Catt. citrina*, Mexico; *Lælia anceps*, Mexico, with [Pg 217] *Epidendrum ciliare*, U.S. Colombia. In other genera there are several hybrids of Mexican and South American parentage; as *L. anceps* × *Epid. ciliare, Sophronitis grandiflora* × *Epid. radicans, Epid. xanthinum* × *Epid. radicans*.

But among Cypripediums, the easiest and safest of all orchids to hybridize, East Indian and American species are unfruitful. Messrs. Veitch obtained such a cross, as they had every reason to believe, in one instance. For sixteen years the plants grew and grew until it was thought they would prove the rule by declining to flower. I wrote to Messrs. Veitch to obtain the latest news. They inform me that one has bloomed at last. It shows no trace of the American strain, and they have satisfied themselves that there was an error in the operation or the record. Again, the capsules secured from very many by-generic crosses have proved, time after time, to contain not a single seed. In other cases the seed was excellent to all appearance, but it has resolutely refused to germinate. And further, certain by-generic seedlings have utterly ignored one parent. *Zygopetalum Mackayi* has been crossed by Mr. Veitch, Mr. Cookson, and others doubtless, with various Odontoglossums, but the flower has always turned out *Zygopetalum Mackayi* pure and simple— [Pg 218]which becomes the more unaccountable more one thinks of it.

Hybrids partake of the nature of both parents, but they incline generally, as in the extreme cases mentioned, to resemble one much more strongly than the other. When a Cattleya or Lœlia of the single-leaf section is crossed with one of the two-leaf, some of the offspring, from the same capsule, show two leaves, others one only; and some show one and two alternately, obeying no rule perceptible to us at present. So it is with the charming *Lælia Maynardii* from *L. Dayana* × *Cattleya dolosa*, just raised by Mr. Sander and named after the Superintendent of his hybridizing operations. *Catt. dolosa* has two leaves, *L. Dayana* one; the product has two and one alternately. Sepals and petals are alike in colour, rosy crimson, veined with a deeper hue; lip brightest crimson-lake, long, broad and flat, curving in handsomely above the column, which is closely depressed after the manner of *Catt. dolosa*.

The first bi-generic cross deserves a paragraph to itself if only on that account; but its own merits are more than sufficient. *Sophro-Cattleya Batemaniana* was raised by Messrs. Veitch from *Sophronitis grandiflora* × *Catt. intermedia*. It flowered in August, 1886; petals and sepals rosy [Pg 219]scarlet, lip pale lilac bordered with amethyst and tipped with rosy purple.

But one natural hybrid has been identified among Dendrobes — the progeny doubtless of *D. crassinode* × *D. Wardianum*. Messrs. J. Laing have a fine specimen of this; it shows the growth of the latter species with the bloom of the former, but enlarged and improved. Several other hybrid crosses are suspected. Of artificial we have not less than fifty.

Phaius — it is often spelt Phajus — is so closely allied with Calanthe that for hybridizing purposes at least there is no distinction. Dominy raised *Ph. irroratus* from *Ph. grandifolius* × *Cal. vestita*; Seden made the same cross, but, using the variety *Cal. v. rubro-occulata*, he obtained *Ph. purpureus*. The success is more interesting because one parent is evergreen, the other, Calanthe, deciduous. On this account probably very few seedlings survive; they show the former habit. Mr. Cookson alone has yet raised a cross between two species of Phajus — *Ph. Cooksoni* from *Ph. Wallichii* × *Ph. tuberculosus*. One may say that this is the best hybrid yet raised, saving *Calanthe Veitchii*, if all merits be considered — stateliness of aspect, freedom in flowering, striking colour, ease of cultivation. One bulb will throw up four spikes — [Pg 220]twenty-eight have been counted in a twelve-inch pot — each bearing perhaps thirty flowers.

Seden has made two crosses of Chysis, both from the exquisite *Ch. bractescens*, one of the loveliest flowers that heaven has granted to this world, but sadly fleeting. Nobody, I believe, has yet been so fortunate as to obtain seed from *Ch. aurea*. This species has the rare privilege of self-fertilization — we may well exclaim, Why! why? —

and it eagerly avails itself thereof so soon as the flower begins to open. Thus, however watchful the hybridizer may be, hitherto he has found the pollen masses melted in hopeless confusion before he can secure them.

One hybrid Epidendrum has been obtained—*Epi. O'Brienianum* from *Epi. evectum* × *Epi. radicans*; the former purple, the latter scarlet, produce ×a bright crimson progeny.

Miltonias show two natural hybrids, and one artificial—*Mil. Bleuiana* from *Mil. vexillaria* × *Mil. Roezlii*; both of these are commonly classed as Odontoglots, and I refer to them elsewhere under that title. M. Bleu and Messrs. Veitch made this cross about the same time, but the seedlings of the former flowered in 1889, of the latter, in 1891. Here we see an illustration of the advantage which French horticulturists enjoy, even so far north as Paris; a clear sky and abundant sunshine made a [Pg 221]difference of more than twelve months. When Italians begin hybridizing, we shall see marvels—and Greeks and Egyptians!

Masdevallias are so attractive to insects, by striking colour, as a rule, and sometimes by strong smell—so very easily fertilized also—that we should expect many natural hybrids in the genus. They are not forthcoming, however. Reichenbach displayed his scientific instinct by suggesting that two species submitted to him might probably be the issue of parents named; since that date Seden has produced both of them from the crosses which Reichenbach indicated.

We have three natural hybrids among Phalœnopsis. *Ph. intermedia* made its appearance in a lot of *Ph. Aphrodite*, imported 1852. M. Porte, a French trader, brought home two in 1861; they were somewhat different, and he gave them his name. Messrs. Low imported several in 1874, one of which, being different again, was called after Mr. Brymer. Three have been found since, always among *Ph. Aphrodite*; the finest known is possessed by Lord Rothschild. That these were natural hybrids could not be doubted; Seden crossed *Ph. Aphrodite* with *Ph. rosea*, and proved it. Our garden hybrids are two: *Ph. F.L. Ames*, obtained from *Ph. amabilis* × *Ph. intermedia*, and *Ph. [Pg 222]Harriettæ* from *Ph. amabilis* × *Ph. violacea*, named after the daughter of Hon. Erastus Corning, of Albany, U.S.A.

Oncidiums yield only two natural hybrids at present, and those uncertain; others are suspected. We have no garden hybrids, I believe, as yet. So it is with Odontoglossums, as has been said, but in the natural state they cross so freely that a large proportion of the species may probably be hybrids. I allude to this hereafter.

I have left Cypripediums to the last, in these hasty notes, because that supremely interesting genus demands more than a record of dry facts. Darwin pointed out that Cypripedium represents the primitive form of orchid. He was acquainted with no links connecting it with the later and more complicated genera; some have been discovered since that day, but it is nevertheless true that "an enormous extinction must have swept away a multitude of intermediate forms, and left this single genus as the record of a former and more simple state of the great orchidacean order." The geographical distribution shows that Cypripedium was more common in early times—to speak vaguely—and covered an area yet more extensive than now. And the process of extermination is still working, as with other primitive types.

[Pg 223]Messrs. Veitch point out that although few genera of plants are scattered so widely over the earth as Cypripedium, the species have withdrawn to narrow areas, often isolated, and remote from their kindred. Some are rare to the degree that we may congratulate ourselves upon the chance which put a few specimens in safety under glass before it was too late, for they seem to have become extinct even in this generation. Messrs. Veitch give a few striking instances. All the plants of *Cyp. Fairieanum* known to exist have sprung from three or four casually imported in 1856. Two bits of *Cyp. superbiens* turned up among a consignment of *Cyp. barbatum*; none have been found since, and it is doubtful whether the species survives in its native home. Only three plants of *Cyp. Marstersianium* have been discovered. They reached Mr. Bull in a miscellaneous case of Cypripediums forwarded to him by the Director of the Botanic Gardens at Buitzenzorze, in Java; but that gentleman and his successors in office have been unable to find another plant. These three must have reached the Gardens by an accident—as they left it—presented perhaps by some Dutchman who had been travelling.

Cyp. purpuratum is almost extinct at Hong [Pg 224] Kong, and is vanishing fast on the mainland. It is still found occasionally in the garden of a peasant, who, we are told, resolutely declines to sell his treasure. This may seem incredible to those who know the Chinaman, but Mr. Roebelin vouches for the fact; it is one more eccentricity to the credit of that people, who had quite enough already. Collectors expect to find a new habitat of *Cyp. purpuratum* in Formosa when they are allowed to explore that realm. Even our native *Cyp. calceolus* has almost disappeared; we get it now from Central Europe, but in several districts where it abounded the supply grows continually less. The same report comes from North America and Japan. Fortunate it is, but not surprising to the thoughtful observer, that this genus grows and multiplies with singular facility when its simple wants are supplied. There is no danger that a species which has been rescued from extinction will perish under human care.

This seems contradictory. How should a plant thrive better under artificial conditions than in the spot where Nature placed it? The reason lies in that archaic character of the Cypriped which Darwin pointed out. Its time has passed—Nature is improving it off the face of the earth. A gradual change of circumstances makes it more and [Pg 225]more difficult for this primitive form of orchid to exist, and, conscious of the fate impending, it gratefully accepts our help.

One cause of extermination is easily grasped. Cypripeds have not the power of fertilizing themselves, except a single species, *Cyp. Schlimii*, which—accordingly, as we may say—is most difficult to import and establish; moreover, it flowers so freely that the seedlings are always weak. In all species the sexual apparatus is so constructed that it cannot be impregnated by accident, and few insects can perform the office. Dr. Hermann Muller studied *Cyp. calceolus* assiduously in this point of view. He observed only five species of insect which fertilize it. *Cyp. calceolus* has perfume and honey, but none of the tropical species offer those attractions. Their colour is not showy. The labellum proves to be rather a trap than a bait. Large insects which creep into it and duly bear away the pollen masses, are caught and held fast by that sticky substance when they try to escape through the lateral passages, which smaller insects are too weak to force their way through.

Natural hybrids occur so rarely, that their existence is commonly denied. The assertion is not quite exact; but when we consider the habits of [Pg 226]the genus, it ceases to be extraordinary that Cypripeds rarely cross in their wild state. Different species of Cattleya, Odontoglots, and the rest live together on the same tree, side by side. But those others dwell apart in the great majority of cases, each species by itself, at a vast distance perhaps from its kindred. The reason for this state of things has been mentioned—natural laws have exterminated them in the spaces between, which are not so well fitted to maintain a doomed race.

Doubtless Cypripeds rarely fertilize—by comparison, that is, of course—in their native homes. The difficulty that insects find in performing that service has been mentioned. Mr. Godseff points out to me a reason far more curious and striking. When a bee displaces the pollen masses of a Cattleya, for instance, they cling to its head or thorax by means of a sticky substance attached to the pollen cases; so, on entering the next flower, it presents the pollen *outwards* to the stigmatic surface. But in the case of a Cypriped there is no such substance, the adhesive side of the pollen itself is turned outward, and it clings to any intruding substance. But this is the fertilizing part. Therefore, an insect which by chance displaces the pollen mass carries it off, as [Pg 227]one may say, the wrong side up. On entering the next flower, it does not commonly present the surface necessary for impregnation, but a sterile globule which is the backing thereof. We may suppose that in the earlier age, when this genus flourished as the later forms of orchid do now, it enjoyed some means of fertilization which have vanished.

Under such disadvantages it is not to be expected that seed capsules would be often found upon imported Cypripeds. Messrs. Veitch state that they rarely observed one among the myriads of plants that have passed through their hands. With some species, however, it is not by any means so uncommon. When Messrs. Thompson, of Clovenfords, bought a quantity of the first *Cyp. Spicerianum* which came upon the market, they found a number of capsules, and sowed them, obtaining several hundred fine plants. Pods are often imported on *Cyp. insigne* full of good seed.

In the circumstances enumerated we have the explanation of an extraordinary fact. Hybrids or natural species of Cypripediums artificially raised are stronger than their parents, and they produce finer flowers. The reason is that they get abundance of food in captivity, and all things are made comfortable for them; whilst Nature, anxious [Pg 228]to be rid of a form of plant no longer approved, starves and neglects them.

The same argument enables us to understand why Cypripeds lend themselves so readily to the hybridizer. Darwin taught us to expect that species which can rarely hope to secure a chance of reproduction will learn to make the process as easy and as sure as the conditions would admit—that none of those scarce opportunities may be lost. And so it proves. Orchidaceans are apt to declare that "everybody" is hybridizing Cypripeds nowadays. At least, so many persons have taken up this agreeable and interesting pursuit that science has lost count of the less striking results. Briefly, the first hybrid Cypripedium was raised by Dominy, in 1869, and named after Mr. Harris, who, as has been said, suggested the operation to him. Seden produced the next in 1874—*Cyp. Sedeni* from *Cyp. Schlimii* × *Cyp. longiflorum*; curious as the single instance yet noted in which seedlings turn out identical, whichever parent furnish the pollen-masses. In every other case they vary when the functions of the parents are exchanged.

For a long time after 1853, when serious work begun, Messrs. Veitch had a monopoly of the business. It is but forty years, therefore, since experiments commenced, in which time hundreds [Pg 229]of hybrids have been added to our list of flowers; but—this is my point—Nature has been busy at the same task for unknown ages, and who can measure the fruits of her industry? I do not offer the remark as an argument; our observations are too few as yet. It may well be urged that if Nature had been thus active, the "natural hybrids" which can be recognized would be much more numerous than they are. I have pointed out that many of the largest genera show very few; many none at all. But is it impossible that the explanation appears to fail only because we cannot yet push it far enough? When the hybridizer causes by force a fruitful union betwixt two genera, he seems to triumph over a botanical law. But suppose the genera themselves are artificial, only links in a grand chain which Nature has forged slowly, patiently, with many a break and many a failure, in the course of ages? She would finish her work bit by bit, and at every stage the new variety may have united with others in endless succession. Few natural hybrids can be identified among Cattleyas, for instance. But suppose Cattleyas are all hybrids, the result of promiscuous intercourse among genera during cycles of time—suppose, that is, the genus itself sprang from parents widely diverse, crossing, [Pg 230]returning, intercrossing from age to age? It is admitted that Cypripedium represents a primeval form—perhaps *the* primeval form—of orchid. Suppose that we behold, in this nineteenth century, a mere epoch, or stage, in the ceaseless evolution? Only an irresponsible amateur could dare talk in this way. It would, in truth, be very futile speculation if experiments already successful did not offer a chance of proof one day, and others, hourly ripening, did not summon us to think.

I may cite, with the utmost brevity, two or three facts which—to me unscientific—appear inexplicable, unless species of orchid were developed on the spot; or the theory of special local creations be admitted. *Oncidium cucullatum* flourishes in certain limited areas of Peru, of Ecuador, of Colombia, and of Venezuela. It is not found in the enormous spaces between, nor are any Oncidiums which might be accepted as its immediate parents. Can we suppose that the winds or the birds carried it over mountain ranges and broad rivers more than two thousand miles, in four several directions, to establish it upon a narrow tract? It is a question of faith; but, for my own part, I could as soon believe that æsthetic emigrants took it with them. But even winds and birds could not bear the seed of *Dendrobium heterocarpum* from Ceylon [Pg 231]to Burmah, and from Burmah to Luzon in the Philippines; at least, I am utterly unable to credit it. If the plants were identical, or nearly, in their different habitats, this case would be less significant. But the *D. heterocarpum* of Ceylon has a long, thin pseudo-bulb, with bright yellow flowers; that of Burmah is short and thick, with paler colouring; that of Luzon is no less than three feet high, exaggerating the stature of its most distant relative while showing the colour of its nearest; but all, absolutely, the same botanic plant. I have already mentioned other cases.

Experience hitherto suggests that we cannot raise Odontoglossum seedlings in this climate; very, very few have ever been obtained. Attempts in France have been rather more successful. Baron Adolf de Rothschild has four different hybrids of Odontoglossum in bud at this present moment in his garden at Armainvilliers, near Paris. M. Moreau has a variety of seedlings.

Authorities admit now that a very great proportion of our Odontoglossums are natural hybrids; so many can be identified beyond the chance of error that the field for speculation has scarcely bounds. O. excellens is certainly descended from O. Pescatorei and O. triumphans, O. elegans from O. cirrhosum and O. Hallii, O. Watti [Pg 232]anum from O. Harryanum and O. hystrix. And it must be observed that we cannot trace pedigree beyond the parents as yet, saving a very, very few cases. But unions have been contracting during cycles of time; doubtless, from the laws of things the orchid is latest born of Nature's children in the world of flora, but mighty venerable by this time, nevertheless. We can identify the mixed offspring of O. crispum Alexandræ paired with O. gloriosum, with O. luteopurpureum, with O. Lindleyanum; these parents dwell side by side, and they could not fail to mingle. We can already trace with assurance a few double crosses, as O. lanceans, the result of an alliance between O. crispum Alexandræ and O. Ruckerianum, which latter is a hybrid of the former with O. gloriosum. When we observe O. Roezlii upon the bank of the River Cauca and O. vexillarium on the higher ground, whilst O. vexillarium superbum lives between, we may confidently attribute its peculiarity of a broad dark blotch upon the lip to the influence of O. Roezlii. So, taking station at Manaos upon the Amazons, we find, to eastward, *Cattleya superba*, to westward *C. Eldorado*, and in the midst *C. Brymeriana*, which, it is safe to assume, represents the union of the two; for that matter, the theory will very soon [Pg 233]be tested, for M. Alfred Bleu has "made the cross" of *C. superba* and *C. Eldorado*, and its flower is expected with no little interest.

These cases, and many more, are palpable. We see a variety in the making at this date. A thousand years hence, or ten thousand, by more distant alliances, by a change of conditions, the variety may well have developed into a species, or, by marriage excursions yet wider, it may have founded a genus.

I have named Mr. Cookson several times; in fact, to discourse of hybridization for amateurs without reference to his astonishing "record" would be grotesque. One Sunday afternoon, ten years ago, he amused himself with investigating the structure of a few Cypripeds, after reading Darwin's book; and he impregnated them. To his astonishment the seed-vessel began to swell, and so did Mr. Cookson's enthusiasm simultaneously. He did not yet know, and, happily, these experiments gave him no reason to suspect, that pseudo-fertilization can be produced, actually, by anything. So intensely susceptible is the stigmatic surface of the Cypriped that a touch excites it furiously. Upon the irritation caused by a bit of leaf, it will go sometimes through all the visible processes of fecundation, the ovary will [Pg 234]swell and ripen, and in due time burst, with every appearance of fertility; but, of course, there is no seed. Beginners, therefore, must not be too sanguine when their bold attempts promise well.

From that day Mr. Cookson gave his leisure to hybridization, with such results as, in short, are known to everybody who takes an interest in orchids. Failures in abundance he had at first, but the proportion has grown less and less until, at this moment, he confidently looks for success in seventy-five per cent. of his attempts; but this does not apply to bi-generic crosses, which hitherto have not engaged his attention much. Beginning with Cypripedium, he has now ninety-four hybrids—very many plants of each—produced from one hundred and forty capsules sown. Of Calanthe, sixteen hybrids from nineteen capsules; of Dendrobium, thirty-six hybrids from forty-one capsules; of Masdevallia, four hybrids from seventeen capsules; of Odontoglossum, none from nine capsules; of Phajus, two from two capsules; of Vanda, none from one capsule; of bi-generic, one from nine capsules. There may be another indeed, but the issue of an alliance so startling, and produced under circumstances so dubious, that Mr. Cookson will not own it until he sees the flower.

It does not fall within the scope of this chapter [Pg 235]to analyze the list of this gentleman's triumphs, but even *savants* will be interested to hear a few of the most remarkable crosses therein, for it is not published. I cite the following haphazard:—

Phajus Wallichii	×	Phajus tuberculosus.
Lœlia præstans.	×	Cattleya Dowiana.
Lœlia purpurata	×	Cattleya Dowiana.
Lœlia purpurata	×	Lœlia grandis tenebrosa.
Lœlia purpurata	×	Cattleya Mendellii.
Lœlia marginata	×	Lœlia elegans Cooksoni.
Cattleya Mendellii	×	Lœlia purpurata.
Cattleya Trianæ	×	Lœlia harpophylla.
Cattleya Percivalliana	×	Lœlia harpophylla
Cattleya Lawrenceana	×	Cattleya Mossiæ.
Cattleya gigas	×	Cattleya Gaskelliana.
Cattleya crispa	×	Cattleya Gaskelliana.
Cattleya Dowiana	×	Cattleya Gaskelliana.
Cattleya Schofieldiana	×	Cattleya gigas imperialis.
Cattleya Leopoldii	×	Cattleya Dowiana.
Cypripedium Stonei	×	Cypripedium Godefroyæ.
Cypripedium Stonei	×	Cypripedium Spicerianum.
Cypripedium Sanderianum	×	Cypripedium Veitchii.
Cypripedium Spicerianum	×	Cypripedium Sanderianum.
Cypripedium Io	×	Cypripedium vexillarium.
Dendrobium nobile nobilus	×	Dendrobium Falconerii.
Dendrobium nobile nobilus	×	Dendrobium nobile Cooksonianum.
Dendrobium Wardianum	×	Dendrobium aureum.
Dendrobium Wardianum	×	Dendrobium Linawianum.
Dendrobium luteolum	×	Dendrobium nobile nobilius.

Masdevallia Tovarensis × Masdevallia bella.
Masdevallia Shuttleworthii × Masdevallia Tovarensis.
Masdevallia Shuttleworthii × Masdevallia rosea.

Of these, and so many more, Mr. Cookson has [Pg 236]at this moment fifteen thousand plants. Since my object is to rouse the attention of amateurs, that they may go and do likewise, I may refer lightly to a consideration which would be out of place under other circumstances. Professional growers of orchids are fond of speculating how much the Wylam collection would realize if judiciously put on the market. I shall not mention the estimates I have heard; it is enough to say they reach many, many thousands of pounds; that the difference between the highest and the lowest represents a handsome fortune. And this great sum has been earned by brains alone, without increase of expenditure, by boldness of initiative, thought, care, and patience; without special knowledge also, at the beginning, for ten years ago Mr. Cookson had no more acquaintance with orchids than is possessed by every gentleman who takes an interest in them, while his gardener the early time was both ignorant and prejudiced. This should encourage enterprise, I think — the revelation of means to earn great wealth in a delightful employment. But amateurs must be quick. Almost every professional grower of orchids is preparing to enter the field. They, however, must needs give the most of their attention to such crosses as may be confi [Pg 237]dently expected to catch the public fancy, as has been said. I advise my readers to be daring, even desperate. It is satisfactory to learn that Mr. Cookson intends to make a study of bigeneric hybridization henceforward. [9]

The common motive for crossing orchids is that, of course, which urges the florist in other realms of botany. He seeks to combine tints, forms, varied peculiarities, in a new shape. Orchids lend themselves to experiment with singular freedom, within certain limits, and their array of colours seems to invite our interference. Taking species and genera all round, yellow dominates, owing to its prevalence in the great family of Oncidium; purples and mauves stand next by reason of their supremacy among the Cattleyas. Green follows — if we admit the whole group of Epidendrums — the great majority of which are not beautiful, however. Of magenta, the rarest of natural hues, we have not a few instances. Crimson, in a thousand shades, is frequent; pure white a little rare, orange much rarer; scarlet very uncommon, and blue almost [Pg 238]unknown, though supremely lovely in the few instances that occur. Thus the temptation to hybridize with the object of exchanging colours is peculiarly strong.

It becomes yet stronger by reason of the delightful uncertainty which attends one's efforts. So far as I have heard or read, no one has yet been able to offer a suggestion of any law which decides the result of combination. In a general way, both parents will be represented in the offspring, but how, to what degree either will dominate, in what parts, colours, or fashions a hybrid will show its mixed lineage, the experienced refuse to conjecture, saving certain easy classes. After choosing parents thoughtfully, with a clear perception of the aim in view, one must "go it blind." Very often the precise effect desired appears in due time; very often something unlooked for turns up; but nearly always the result is beautiful, whether or no it serve the operator's purpose. Besides effect, however, there is an utility in hybridization which relates to culture. Thus, for example, the lovely *Cypripedium Fairieanum* is so difficult to grow that few dealers keep it in their stock; by crossing it with *Cyp. barbatum*, from Mount Ophir, a rough-and-ready cool species, we get *Cyp. vexil* [Pg 239]*larium*, which takes after the latter in constitution while retaining much of the beauty of the former. Or again, *Cypripedium Sanderianum*, from the Malay Archipelago, needs such swampy heat as few even of its fellows appreciate; it has been crossed with *Cyp. insigne*, which will flourish anywhere, and though the seedlings have not yet bloomed, there is no reasonable doubt that they will prove as useful and beautiful as in the other case. *Cypripedium insigne*, of the fine varieties, has been employed in a multitude of such instances. There is the striking *Cyp. hirsutissimum*, with sepals of a nameless green, shaded yellow, studded with spiculæ, exquisitely frilled, and tipped, by a contrast almost startling, with pale purple. It is very "hot" in the first place, and, in the second, its appearance would be still more effective if some white could be introduced; present it to *Cyp. niveum* and confidently expect that the progeny will bear cooler treatment, whilst their "dorsal sepal" will be blanched. So the charming *Masdevallia Tovarensis*, warm, white and lowly, will take to itself the qualities, in combination, of *Mas. bella*, tall, cool, and highly coloured red and yellow, as Mr. Cookson has proved; so *Phalœnopsis Wightii*, delicate of growth and small of flower, will become strong [Pg 240]and generous by union with *Phal. grandiflora*, without losing its dainty tones.

It is worth mention that the first Flora medal offered by the Royal Horticultural Society for a seedling—a hybrid—in open competition was won by *Lœlia Arnoldiana* in 1891; the same variety took the first prize in 1892. It was raised by Messrs. Sander from *L. purpurata* × *Catt. labiata*; seed sown 1881, flowered 1891.

And now for the actual process by which these most desirable results, and ten thousand others, may be obtained. I shall not speak upon my own authority, which the universe has no reason to trust. Let us observe the methods practised in the great establishment of Mr. Sander at St. Albans.

Remark, in the first place, the low, unshaded range of houses devoted to hybridization, a contrast to those lofty structures, a hundred yards long or more, where plants merely flourish and bloom. Their span roofs one may touch with the hand, and their glass is always newly cleaned. The first and last demand of the hybridizer is light—light—eternally light. Want of it stands at the bottom of all his disappointments, perhaps. The very great majority of orchids, such as I refer to, have their home in the tropics; even the "cool" Odontoglots and Masde [Pg 241]vallias owe that quality to their mountaineering habit, not to latitude. They live so near the equator that sunshine descends almost perpendicularly—and the sun shines for more than half the year. But in this happy isle of ours, upon the very brightest day of midsummer, its rays fall at an angle of 28°, declining constantly until, at midwinter, they struggle through the fogs at an inclination of 75°. The reader may work out this proportion for himself, but he must add to his reckoning the thickness of our atmosphere at its best, and the awful number of cloudy days. We cannot spare one particle of light. The ripening seed must stand close beneath the glass, and however fierce the sunshine no blind may be interposed. It is likely that the mother-plant will be burnt up—quite certain that it will be much injured.

This house is devoted to the hybridizing of Cypripediums; I choose that genus for our demonstration, because, as has been said, it is so very easy and so certain that an intelligent girl mastered all its eccentricities of structure after a single lesson, which made her equally proficient in those of Dendrobes, Oncidiums, Odontoglots, Epidendrums, and I know not how many more. The leaves are green and smooth as yet, with many a fantastic bloom, and many an ovary that has just [Pg 242]begun to swell, rising amidst the verdure. Each flower spike which has been crossed carries its neat label, registering the father's name and the date of union.

Mr. Maynard takes the two first virgin blooms to hand: *Cypripedium Sanderianum*, and *Cypripedium Godefroyæ*, as it chances. Let us cut off the lip in order to see more clearly. Looking down now upon the flower, we mark two wings, the petals, which stood on either side of the vanished lip. From the junction of these wings issues a round stalk, about one quarter of an inch long, and slightly hairy, called the "column." It widens out at the tip, forming a pretty table, rather more than one-third of an inch long and wide. This table serves no purpose in our inquiry; it obstructs the view, and we will remove it; but the reader understands, of course, that these amputations cannot be performed when business is intended. Now—the table snipped off—we see those practical parts of the flower that interest us. Beneath its protection, the column divides into three knobbly excrescences, the central plain, those on either side of it curling back and down, each bearing at its extremity a pad, the size of a small pin's head, outlined distinctly with a brown colour. It is quite impossible to mistake these things; equally [Pg 243]impossible, I hope, to misunderstand my description. The pads are the male, the active organs.

But the column does not finish here. It trends downward, behind and below the pads, and widens out, with an exquisitely graceful curve, into a disc one-quarter of an inch broad. This is the female, the receptive part; but here we see the peculiarity of orchid structure. For the upper surface of the disc is not susceptible; it is the under surface which must be impregnated, though the imagination cannot conceive a mere accident which would throw those fertilizing pads upon their destined receptacle. They are loosely attached and adhesive, when separated, to a degree actually astonishing, as is the disc itself; but if it were possible to displace them by shaking, they could never fall where they ought. Some outside impulse is needed to bring the parts together. In their native home insects perform that service—sometimes. Here we may take the first implement at hand, a knife, a bit of stick, a pencil. We remove the pads, which yield at a touch, and cling to the object. We lay them one by one on the receptive disc, where they seem to melt into the surface—and the trick is done. Write out your label—"*Cyp. Sanderianum* × *Cyp. Godefroyæ*, Maynard." Add the date, and leave Nature to her work.

[Pg 244]She does not linger. One may almost say that the disc begins to swell instantly. That part which we term the column is the termination of the seed-purse, the ovary, which occupies an inch, or two, or three, of the stalk, behind the flower. In a very few days its thickening becomes perceptible. The unimpregnated bloom falls off at its appointed date, as everybody knows; but if fertilized it remains entire, saving the labellum, until the seed is ripe, perhaps half a year afterwards—but withered, of course. Very singular and quite inexplicable are the developments that arise in different genera, or even species, after fertilization. In the Warscewiczellas, for example, not the seed-purse only, but the whole column swells. *Phalænopsis Luddemanniana* is specially remarkable. Its exquisite bars and mottlings of rose, brown, and purple begin to take a greenish hue forthwith. A few days later, the lip jerks itself off with a sudden movement, as observers declare. Then the sepals and petals remaining take flesh, thicken and thicken, while the hues fade and the green encroaches, until, presently, they assume the likeness of a flower, abnormal in shape but perfect, of dense green wax.

This Cypripedium of ours will ripen its seed in about twelve months, more or less. Then [Pg 245]the capsule, two inches long and two-thirds of an inch diameter, will burst. Mr. Maynard will cut it off, open it wide, and scatter the thousands of seeds therein, perhaps 150,000, over pots in which orchids are growing. After experiments innumerable, this has been found the best course. The particles, no bigger than a grain of dust, begin to swell at once, reach the size of a mustard-seed, and in five or six weeks—or as many months—they put out a tiny leaf, then a tiny root, presently another leaf, and in four or five years we may look for the hybridized flower. Long before, naturally, they have been established in their own pots.

Strange incidents occur continually in this pursuit, as may be believed. Nine years since, Mr. Godseff crossed *Catasetum macrocarpum* with *Catasetum callosum*. The seed ripened, and in due time it was sown; but none ever germinated in the proper place. A long while afterwards Mr. Godseff remarked a tiny little green speck in a crevice above the door of this same house. It grew and grew very fast, never receiving water unless by the rarest accident, until those experts could identify a healthy young Catasetum. And there it has flourished ever since, receiving no attention; for it is the first rule in orchid culture to leave a plant [Pg 246]to itself where it is doing well, no matter how strange the circumstances may appear to us. This Catasetum, wafted by the wind, when the seed was sown, found conditions suitable where it lighted, and quickened, whilst all its fellows, carefully provided for, died without a sign. It thrives upon the moisture of the house. In a very few years it will flower. In another case, when all hope of the germination of a quantity of seed had long been lost, it became necessary to take up the wooden trellis that formed the flooring of the path; a fine crop of young hybrids was discovered clinging to the under side.

The amateur who has followed us thus far with interest, may inquire how long it will be before he can reasonably expect to see the outcome of our proceedings? In the first place, it must be noted that the time shortens continually as we gain experience. The statements following I leave unaltered, because they are given by Messrs. Veitch, our oldest authority, in the last edition of their book. But at the Temple Show this year Norman C. Cookson, Esq., exhibited *Catt. William Murray*, offspring of *Catt. Mendellii x Catt. Lawrenceana*, a lovely flower which gained a first class certificate. It was only four years old.

The quickest record as yet is *Calanthe Alex [Pg 247]anderii*, with which Mr. Cookson won a first-class certificate of the Royal Horticultural Society. It flowered within three years of fertilizing. As a genus, perhaps, Dendrobiums are readiest to show. Plants have actually been "pricked out" within two months of sowing, and they have bloomed within the fourth year. Phajus and Calanthe rank next for rapid development. Masdevallia, Chysis, and Cypripedium require four to five years, Lycaste seven to eight, Lœlia and Cattleya ten to twelve. These are Mr. Veitch's calculations in a rough way, but there are endless exceptions, of course. Thus his *Lœlia triophthalma* flowered in its eighth season, whilst his *Lœlia caloglossa* delayed till its nineteenth. The genus *Zygopetalum*, which plays odd tricks in hybridizing, as I have mentioned, is curious in this matter also. *Z. maxillare* crossed with *Z. Mackayi* demands five years to bloom, but *vice versâ* nine years. There is a case somewhat similar, however, among the Cypripeds. *C. Schlimii* crossed with *C. longifolium* flowers in four years, but *vice versâ* in six. It is not to be disputed, therefore, that the hybridizer's reward is rather slow in coming; the more earnestly should he take measures to ensure, so far as is possible, that it be worth waiting for.

FOOTNOTES:

[9] Mr. Cookson writes to me: "Give some of the credit to my present gardener, William Murray, who is entitled to a large proportion, at least."

[Pg 248]

www.ingramcontent.com/pod-product-compliance
Lightning Source LLC
Chambersburg PA
CBHW031416210526
45464CB00005B/1906